高忠義 —— 著

律師教你寫
英文契約

專業人士必備的實務指南

CONTRACT

Contents

第三章

給付義務句

第四章

裁量選擇句與宣示句

第七章

英文契約的核心事項條款組合

第八章

附隨義務條款組合與保密條款

第九章

競業禁止條款與獨家交易條款

第十章

彌償條款與智慧財產權條款

第十一章 ────────────────────────

契約的轉折部分，與情事變更相關條款

第十二章 ────────────────────────

違約救濟條款

第十三章

散場條款

第十四章

紛爭解決條款

第一章

導論

作者從事法律工作二十多年，深深體會契約審閱與撰擬的重要性，尤其在臺商全球化的佈局與經營之下，法律工作者更需要對英文契約徹底清楚地掌握，以避免因為簽訂的契約不當使企業面對預料之外的風險。

許多企業法務，需要審閱或撰擬英文契約時，可能就是在網路上搜尋範本或資料，東抄西剪，寫出來的東西可能未切中要點。

坊間大部分關於英文契約的著作，也只有拼貼各種常見的語句、用詞，讓讀者知其然而不知其所以然地複製套用，並未建立契約撰擬的知識系統，許多甚至崇尚推廣彆扭難懂的法律業界慣用語詞，無助於溝通，反而製造誤會與糾紛。

因此，作者利用在臺美兩地學習的成果，以及二十多年累積的工作經驗，以「445」這樣易懂易記的數字組合彙整契約撰擬的知識系統，說明契約治理的四種模式，契約的四個重點內容，以及撰擬契約的五個面向，方便讀者全盤掌握模組化、系統化的契約撰擬知識，而能運用在各種類型的交易之上。

第一節　契約治理的四種模式

2009年諾貝爾經濟學獎得主Oliver E. Williamson提出的契約治理模型，按照交易頻率高與低，以及專屬投資的程度，對交易進行分類（詳如表1），並針對不同類型的交易建議採用不同的契約關係治理模式（詳如表2）。

表1　契約關係分類

	無專屬投資	專屬投資程度居中	需要高度專屬投資
偶爾交易	採購標準化的設備	採購客製化設備	建造一座工廠
經常交易	採購標準化的材料	採購客製化材料	不同廠址連續加工

交易頻率，指是否經常進行這樣的交易。而專屬投資的程度，指是否需要為了某個交易而進行這個交易專用的投資，投資得到的成果，無法轉用在別的客戶的交易上，也無法轉賣相關資產，所以如果交易獲利未符預期，做出的專屬投資就打水漂了。

Williamson在上述的分類基礎上針對不同的交易類型建議採用下表所述的治理模式：

表2　契約治理模式

	無專屬投資	專屬投資程度居中	需要高度專屬投資
偶爾交易	市場治理	三方治理	
經常交易		雙方治理	一體化治理

對於無專屬投資的交易，無論交易頻率高或低，均建議採用市場治理模式，這樣的產品有許多供應商，買方可以很容易地找到替代產品，市場力量可以自然而然地約束調整。

對於需要專屬投資，而且交易頻率不高的情況，Williamson建議採用三方治理的模式。例如：屋主找人來裝修房子，相關的設計、用料、以及施工完成後的成果，都不太可能轉用到別的案件上。這種交易適合採用三方治理模式，也就是找第三人來協調、監督整個施工的過程，例如：較大的案子會找監工。

　　如果是需要專屬投資，而且交易頻率很高的情況，Williamson建議採用一體化的治理模式。舉個作者看到的例子，Youbike的腳踏車是某家知名自行車品牌廠商提供的，但是停放Youbike的停車架跟各項靠卡扣款的軟硬體則是由另一家資訊公司在維持。因此，自行車品牌廠商跟資訊公司必須經常地往來，而且資訊公司也要配合自行車品牌廠商配置專責人員，提供各種量身訂作的服務。如果自行車品牌廠商每一次資訊公司處理某個地點停車架或軟硬體問題，都必須互相商量人員工時費用，完修時間，那肯定是很沒效率的。因此，為了Youbike的專案，自行車品牌廠商與資訊公司必須走向一體化的治理。他們可能一起合作跟政府投標，得標之後再分潤，而且彼此之間可能有交互持股投資的整合行為，才更能利害一體，休戚與共。當然一體化的極致作法，就是兩家公司併作一家，將外部關係內部化。

　　至於專屬投資屬於適中情況，而雙方當事人需要經常往來交易的情況，則是採取雙方治理的模式。雙方需要針對特定事項相互協調，但另外一些事項可能保留自主彈性空間。(註1)

　　有個例子可以生動地解釋這四種治理模式。男女交往的發展過程常常就是這四種歷程。一開始可能是一群男男女女一起參加大型的聯誼活動，如果有哪一對看對眼，想要更進一步認識，也許找個媒人、介紹人或閨密之類的權充電燈泡，等到兩個人互相熟悉了之後，就自行帶開，到最後濃情密意，願意攜手共度一生，也許就登記結婚了。

1　上文提及Williamson的部分，見Oliver E. Williamson, Transaction-Cost Economics: The Governance of Contractual Relation, Journal of Law and Economics, Vol. 22, No. 2 (Oct., 1979), pp. 233-261, The University of Chicago Press。表1簡化自Ibid., p. 247, Figure 1。表2簡化自Ibid., p. 253, Figure 2。

第二節　契約的四個重點內容

　　作者小時候喜歡看布袋戲，因此用布袋戲的情節套路來說明契約的四個重點內容。

1. 來者何人，報上名來

　　主角在江湖路上行走之時，必然是要跟另一角色狹路相逢，這時主角就會說「來者何人，報上名來」。

　　契約的第一個重點內容，就是來者何人，報上名來。作者曾經處理過一家臺商的案子，這個臺商跟美國客戶做生意做了二十多年，突然拖欠款項，雙方之間未曾簽訂正式的書面契約，都是訂單往來，回頭翻了訂單，用訂單上列印的公司名字Adam找尋美國客戶所在該州的公司登記資料庫，發現同一地址有多家公司的名字都是「Adam」開頭。Adam U.S.A Ltd.、Adam International. Inc.、Adam Trading LLC.，即使翻找匯款記錄，每一次付款的公司名字也有不同。因此不論是要催告或訴訟，都無從確定往來的客戶是究竟是哪一家Adam公司。這就是沒有掌握契約第一個重點的慘痛教訓。

　　在法律上能夠簽約當主體，能夠承擔義務、享受權利的就只有「人」，而所謂「人」又有兩種，「自然人」與「法人」。「自然人」指的就是活生生的每一個人。而「法人」則是具有法律上人格的組織體，又可分為「社團法人」與「財團法人」。「社團法人」是人的結合，所以重視成員的權益。例如：公司或協會之類。而「財團法人」則是用特定金錢或財產為特定公益目的使用而登記成組織體，所以一般口語講的財團常常指的是好多家公司形成一個集團，但法律上的財團法人指的卻是

為了公益目的而捐出一筆錢或特定資產，做特定用途，例如：基金會，不可不辨。

　　生活中我們常常發現有些契約記載的當事人卻是「診所」、「藥局」、「企業社」、「商號」、「工廠」、「合夥」甚至很多也有統一編號、稅務編號。但這些都不是法人，因此在簽訂契約時，務必要找出這些機構幕後的藏鏡人，也就是負責人。

　　簽訂契約時，對於交易對象的掌握必須非常清楚而明白，當事人的資訊通常出現在契約標題之下的第一段，其具體內容將在本書後續章節中再詳細說明。

2. 有何指教

　　主角與另一角色狹路相逢，互通名姓之後，往往就是探詢有何指教，要吃飯喝酒，問事探路。契約也是如此，締約必然有目的，其目的用最白話的方式來說就是「交換」。吃飯喝酒，問事探路都不需要簽契約，能簽約的就是交換。英美法系強調契約必須有約因（consideration），也就是要有對價，要相互交換利益，這種利益不需要對稱或相當，只要當事人甘心樂意就好。臺灣採用的歐陸法系制度雖然不要求約因，但像是贈與契約，在交付贈與物之前原則上仍可自由撤銷，所以結果可以說是有些相近，一方承諾給予他方好處，他方不給任何回報，這種合意很難上法院強制要求履行。

3. 千里相逢，終須一別

　　徐志摩的詩作〈偶然〉裡有這麼一句「你我相逢在黑夜的海上，你有你的，我有我的，方向」。兩人之相逢或喜或悲，終有一別。契約也是如此，雙方的交換無論結果是開心地完成，亦或是不成而失望，但終有分別之時，各奔一方。契約必須妥慎處理分別之事，最痛苦的事情是不能好好說再見，無休無止的拉扯撕逼。舉例來說，賣方有個房子要賣，結果買方遲遲不匯入價金，賣方如果未解除契約就將房子再轉賣給第三人，這就變成賣方違約了，買方可能反咬賣方違約造成給付不能，因此違約賠償是一回事，但是交易確定卡死不能做的時候，必須將契約關係消滅掉，否則也會卡住自己的退路。

　　契約的交換若順利圓滿完成，瑕疵擔保或保固期間也都完了，那自然是皆大歡喜，不須贅言。但如果交換過程中有任何摩擦，則須訴諸終止（termination）、解除（repudiation）、撤銷（rescission）等機制讓契約關係作清楚的結束，以利相互結清未了責任。這三種機制將在本書後續有關契約具體條款中詳予說明。

4. 到哪裡輸贏

　　布袋戲的劇情幾乎都是決鬥打架，決鬥打架常常要約個時間、地點，甚至有時還要有公親在場見證，避免一方用詭詐之術暗算。契約交換發生糾紛也是常有之事，這樣的糾紛也需要有個處理方式；要用訴訟、仲裁，還是調解，要約定採用什麼樣的程序，採用哪一種準據法。

第三節　契約撰擬的五個面向

以蓋房子做譬喻。蓋房子需要有**材料**，例如：磚瓦、木泥；也需要確立**結構**，有基礎、有建築主體、有外部的包覆跟內部的樑柱、牆壁、天花板、地板等等；而房子內部也會按照需求的**功能**進行隔間，例如：客廳、臥室、廚房、衛浴等等；蓋房子也必然有一定的**工作順序**，例如：整地，備料、準備機具、一層一層地往上蓋；最後蓋房子不能只考慮屋主有哪些需求，想要蓋出什麼樣的房子，還需要考慮外部法規有什麼規範，蓋的過程不能損害鄰地鄰屋，有都市計畫分區、容積率、建蔽率，消防安全的各種規範。這些規範都是控管外部性，避免蓋房子的人太自私只想到自己，沒有想到別人受到各種損害，而要求遵循控管負面外部性的**外部規範**。

寫契約與蓋房子很像，以作者的工作經驗心得整理出契約撰擬的五個面向：句法、結構、功能目的、時序、外部規範。

作者在美國西北大學研讀碩士時特別選了契約撰擬的課程，也因為對於契約撰擬極有興趣，所以除了上課之外，還找了許多書來參考，從而瞭解在美國的契約操作實務中，契約每一個句子都可以歸屬到某個類型，而且契約從標題到最後簽名，也幾乎有固定的架構。

契約是有法律效力的文件，是安排規劃雙方交換各種價值過程的應用文書。也因為契約是有功能、有目的之文件，要安排的事情很多，所以我們必須字斟句酌追求精煉，不要浪費筆墨在廢話、無用的詞語，讓雙方都更加耗神，反而容易遺誤重要之事。契約的每一句話都有想要達成的法律效果，也因此可以從想要達成的效果對契約句型進行分類，作者提出「天地陰陽」的概念作為分類方法，下一章我們將詳細說明。

　　同樣地，在美國的實務操作下，契約從標題到簽名也幾乎已經有固定的結構，我們可以套用小說故事的「起承轉合」加以說明，這部分也將在後續有關具體條款的章節中說明。

　　如前所述，契約的核心功能在於「交換」，而「交換」必然有其標的對象。以作者的實務工作經驗來濃縮整理。可以說契約只有五種交換標的：人流、物流、金流、資訊流與權利的流通。人流就是以人做什麼事或不做什麼事為交換標的；物流就是以實體的物作為交換標的；金流就是以金錢作為交換標的；資訊流就是以資訊作為交換標的；權利流就是以各種權利作為交換標的。而這些權利可能連結到特定物品，或者並無連結的特定物品。

　　契約的核心功能在於進行這五種交換，而契約要填上的內容就是圍繞著這五種交換的各項具體工作安排，例如：在長期供應契約中，大致的程序是賣方先發送樣品或規格資訊給買方，買方測試樣品或確認規格並通知賣方，買方提供需求預測資料給賣方，賣方進行備料、買方發送訂單，賣方確認，賣方交貨，買方驗收付款。

　　契約五大面向中的最後一個面向則是遵循外部規範，例如：外國自然人或法人有意投資臺灣公司，在某些情況下必須向經濟部投審會申請，例如：某些電子零組件、機械設備、或化學產品的進出口受到輸出管制法規的限制，因為美國政府設了一些黑名單，擔心這些產品的買方拿來做大規模毀滅武器，而且強制要求世界各國都遵守。又像是藥品、醫療器材、化妝品也有程度或寬或緊的各種管制規範，不是買賣雙方條件談妥就能隨意買賣。

英文契約基本句型介紹，
聲明句與擔保句

第一節　英文契約基本句型介紹

　　美國的契約實務已發展出模組化的寫作方式。就像蓋房子那樣，我們利用各種材料（磚瓦、石材、塑料、鐵片、木材、水泥、鋼筋）建造出有基礎、主體、門窗、隔間、外牆的房子。寫契約也是如此，我們可以利用各種句型的句子（宣示句、聲明句、擔保句、給付義務句、裁量選擇句、履行句等）完成契約的前言、敘文（背景）、定義、主要行動、契約終結事項與一般條款。

　　關於英文契約有哪些基本句型，許多人提出不同的看法，有人說是「宣示（declarations）、聲明（representations）、擔保（warranties）、給付義務（covenants）、裁量選擇（discretionary authority）、條件（conditions）」六種[註1]。也有人主張分成「合意（agreement）、即時履行（performance）、義務（obligation）、裁量（discretion）、禁止（prohibition）、政策（policy）、宣示（declaration）、條件（conditions）、信念（belief）、意圖（intention）、建議（recommendation）」等十種契約語言[註2]。句型分類的重點在於實用，容易操作，無法用理論辨證孰是孰非。而區分種類更多的方法，有時也只是將某個大類劃分得更細。作者較傾向於採用前述第一種分類方法，但加上第二種分類方法的即時履行（performance），合為七種。[註3]

　　這七種基本句型本身並不容易記住，因此作者獨創太極圖式的理解

1　Tina L. Stark, *Drafting Contracts: How and Why Lawyers Do What They Do* 9（Wolters Kluwer, 2007）.

2　Kenneth A. Adams, *A Manual of Style for Contract Drafting* 39-98（3rd ed., American Bar Association, 2013）.

3　本書作者認為，信念、意圖、建議並不發生法律上的效力，因此不宜作為契約句型。禁止事項可歸併為給付義務中的不作為義務，較為簡便。合意文句（agreement）其實是契約結構中的一個要項，並不是契約句型。所謂契約句型，指的是在整份契約中頻繁使用的構句方法，合意文句只出現在背景介紹之後的一句話，並不是反覆使用的構句方法。而政策句型可併入宣示句較為簡潔。

方法。以「天地陰陽」區分這七種基本句型，「天」指將來的事，例如：給付義務，也就是當事人在未來某個時間點有義務做什麼事。「地」指已發生的事，例如：聲明已具有何種資格。「陽」指剛性安排，如果違反，即有違約責任，例如：給付義務，違反可能有賠償責任。「陰」指柔性安排，約定即生效果，無違反可能性，例如：裁量選擇。詳如下圖及各節逐項說明。

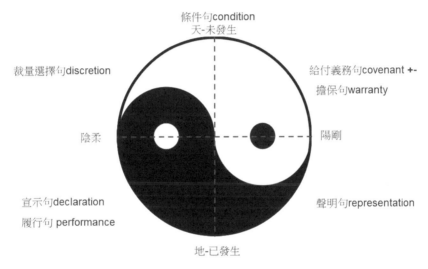

圖 1　契約基本句型

我們要特別說明，在本書中提到句型，指的是造句方式，提到條款是指契約中一些固定常見的條款，例如：付款條款，交貨條款。而條款中往往會有多個句子，每一個句子可能分屬不同句型。例如：付款條款，通常以給付義務句型為核心，表述一方當事人應在某年某月某日支付特定金額的特定貨幣。但契約中往往有聲明條款與擔保條款。聲明條款與擔保條款的核心句型是聲明句與擔保句。所以請務必分清楚條款與句型的差異。

第二節　聲明句

案例一

Bisset v Wilkinson [1927] AC 177 [註4]

　　1919年5月，<u>畢塞特</u>（Bisset）簽了一份契約將兩塊相鄰的農地以13260英鎊的價格銷售給<u>威金森</u>（Wilkinson）。這兩塊地分別是834公頃與141公頃。在議約過程中，<u>畢塞特</u>告訴<u>威金森</u>，「只要有素質好的六人騎士團隊，這個農場應該可以養2,000隻羊」。兩年經營不順之後，<u>威金森</u>認定這片地沒辦法養2,000隻羊，他也提起不實聲明之訴，請求解除契約退款。

　　法院認定前述有關農場情況的說明，並不是基於<u>畢塞特</u>所知的嚴肅聲明。在交易當時，雙方當事人都瞭解<u>畢塞特</u>並未用那塊地來放羊，因此有關農地能養幾隻羊也只是大概的估算。法院也表示重要的考量點在於「交易的重要事實、雙方當事人個別所知情況，與他們的相對地位，聲明的具體文字，以及談論的主旨事項之實際情況……」也補充提到：在確認有關能養兩千隻羊的說明究竟傳達給原告什麼意思時，最重要該記得的事情是，雙方當事人都知道，被告並未用出售的土地來放羊，就雙方所知，也沒有任何人用出售的地來養過羊。而且原告也未證明如果那個農地妥慎管理，真得養不了兩千隻羊。

4　https://en.wikipedia.org/wiki/Bisset_v_Wilkinson，最後點閱日，May 7，2023。

案例二

Esso Petroleum v Mardon - 1976 [註5]

　　麥頓（Mardon）加盟經營埃索石油（Esso Petroleum）加油站。在議約過程中，埃索表示那個加油站一年預估可以賣出20萬加侖的石油，但是當地議會在區域規劃上做出某項決定，可能讓車輛無法從主要幹道通往那個加油站，因此加油站的客戶數量應該會減少。埃索提供的估算並未考量這項因素。麥頓簽約加盟，但營業狀況不理想。之後，麥頓跟埃索石油協商談好減租，但是還是虧損。埃索後來起訴要求終止加盟契約，麥頓反訴主張埃索違反擔保，或者有過失。原審判決不利於麥頓，麥頓上訴。

　　上訴法院認定本案並不是擔保，埃索確實並未擔保一年可以賣出20萬加侖。但是那確實是埃索的估算，而埃索有專門知識與技術。那就是衡量加油站的指標。埃索知道事實。他們知道城裡的交通狀況。他們知道類似加油站的銷量。他們有豐富經驗與專業能力可用。他們比麥頓有更好地位可以做預測。如果有這樣地位的人做了預測，希望藉此讓另一方據以行事，而他方確實據此行事，有充分理由可以解釋成當事人一方擔保在預估時盡了合理的注意也善用其技術，所以那樣的預估是恰當的，而且是可靠的。埃索石油向麥頓説，我們預估銷量是20萬加侖。你可以信賴對這個加油站來説，這是恰當的預估數字。租金也是按照這個預估數字推算的。但結果顯示這不是恰當的預估數字，有技術有經驗的人不會這樣推估，這就構成違反擔保。在議約時有注意義務。如果一方確實具有或宣稱具有特殊知識與技術，作出聲明，希望以此誘使他方締約，而他方也確實因此願意締約，那麼聲明

5　https://en.wikipedia.org/wiki/Esso_Petroleum_Co_Ltd_v_Mardon#Judgment，最後點閱日，May 9，2023。

> 的一方就有合理的注意義務必須運用適當的注意義務以確保他的
> 聲明是正確的，而且他提出的建議、資訊或意見是可靠的。埃索
> 確實聲稱有專業知識，而確實也有。他們過失的不實聲明是嚴重
> 的錯誤。無論提供建議有沒有報酬作為對價，都負擔相同的注意
> 義務。只要違反這樣的義務就應負損害賠償責任，而且不管是基
> 於契約起訴，或基於侵權而起訴，都是一樣的。

1. 聲明的目的與聲明不實的法律責任

契約中的聲明條款就是一方當事人用來陳述重要事實以吸引另一方
當事人願意締約。如果一方當事人說假話騙人，或隱匿重要事實，或用
其他方式誤導別人，可能造成他方誤認重要事實，締結契約之後才發現
情況完全跟預想的不一樣，而蒙受重大損害。

舉例來說，買了房子，之後才發現是凶宅，海砂屋、輻射屋等各種
情況。在企業併購中更是如此，賣方土地裡面有沒有未清乾淨的污染、
賣方的智慧財產權有沒有被舉發、異議，賣方的產品有沒有侵害第三人
的智慧財產權、賣方有沒有未了的訴訟、賣方有沒有積欠勞工薪水、賣
方的資產或存貨有沒有抵押給第三人、賣方有沒有替第三人背書保證、
賣方有多少特別股、可轉債或認股權證流通在外、賣方有沒有積欠稅款
等，這些問題如果締約時認識不清，事後才知道就很頭痛了。

美國在十九世紀時，開始快速由「好價錢就必須買到好商品」（a
sound price warrants a sound commodity）的原則轉向支持「買方自
己當心」（caveat emptor）的原則，法院不再像以前那樣替當事人計較
他們的交易是不是方方面面都很公道。原因之一，在於美國的經濟正在

快速發展、各種新把戲、新玩意都出現了，不像過去的世界那樣單純，法官很難找出客觀的標準進行干預（註6）。再者美國人有很強烈的自治傾向，不希望政府法律干預私人間生意上的事。而且新經濟、新技術往往走在法律前頭，法官一直扯後腿，社會也不容易進步。當然目前又不太一樣了，公權力介入在許多時候又被認為是必要的，尤其在大企業與消費者的交易關係上。但是在B2B或C2C的契約中，還是要當事人自己把話說清楚。

　　法律可能對說謊騙人的人課予某些法律責任，以避免人們被騙損失。臺灣刑法中第339條（註7）有刑事上的詐欺罪，臺灣民法第92條（註8）也規定被詐欺時可以撤銷意思表示。另外被詐欺的人在臺灣也可能依據民法第184條（註9）請求侵權的損害賠償。

　　在美國，美國法律學會（American Law Institute）編纂的法律整編（Restatement of Law）彙整了common law的原則，在契約法第二次整編（Restatement〔Second〕of Contracts）第七章主題一，與侵權法第二次整編第二十二章都對不實陳述的類型、定義與效果做了說明。

　　契約法第二次整編第七章主題一的規範著重在什麼樣的作為構成不實聲明，重要條款包括不符合事實的宣稱為不實聲明（第159條），或

6　參閱 Morton J. Horwitz, *Historical Foundations of Modern Contract Law*, 87 Harv.L.Rev. 945-946（1974）轉引自 Stephen B. Presser & Jamil S. Zainaldin, *Law and Jurisprudence in American History*, 320-21（6ed. 2005）。

7　「1.意圖為自己或第三人不法之所有，以詐術使人將本人或第三人之物交付者，處五年以下有期徒刑、拘役或科或併科五十萬元以下罰金。2.以前項方法得財產上不法之利益或使第三人得之者，亦同。3.前二項之未遂犯罰之。」

8　「1.因被詐欺或被脅迫而為意思表示者，表意人得撤銷其意思表示。但詐欺係由第三人所為者，以相對人明知其事實或可得而知者為限，始得撤銷之。2.被詐欺而為意思表示，其撤銷不得以之對抗善意第三人。」

9　「1.因故意或過失，不法侵害他人之權利者，負損害賠償責任。故意以背於善良風俗之方法，加損害於他人者亦同。2.違反保護他人之法律，致生損害於他人者，負賠償責任。但能證明其行為無過失者，不在此限。」

者以行為阻撓他方知悉某項事實而形同主張該事實不存在的宣稱（第160條），或者在某些情況下，不主動揭露某些資訊，也形同主張該事實不存在的宣稱（第161條）。不實聲明怎樣才構成詐欺，或有重大性（第162條）。有些情況，因為不實聲明，所以認定契約根本不成立（第163條）。有些情況，因為不實聲明，契約雖然是成立了，但可撤銷（第164條）。有些情況，受不實聲明影響的當事人，可以聲請修改契約（reformation），而不是只能撤銷契約（第166條）。某些情況，意見的陳述可以推論意見陳述人知道某些事實，而且有正當理由提出那樣的意見，所以雖然只是提出意見也要負責（第167條）

侵權法第二整編第二十二章則將不實聲明分為三類，其一是詐欺的（fraudulent）不實聲明、其二是過失的（negligent）不實聲明、其三是無過失的（innocent）不實聲明。如果是詐欺的不實聲明，可以請求金錢損失（pecuniary loss），計算方法是（a）收到給付的實際價值與其購買的金錢價格或其他對待給付價值的差異。（b）因為信賴不實聲明而蒙受的其他金錢損失。（c）有些情況也可以請求移轉不實聲明人因此獲得的不當得利。（第549條）過失不實聲明的則排除前述（c）的不當得利移轉請求。（第552B條）若是無過失的不實聲明，則只能請求賠償價差，而且僅限於買賣、租賃或互易的契約。（第552C條）

在某些情況下，法院也會許可侵權的懲罰性損害賠償，但美國各州法院對什麼情況下可以針對詐欺的不實聲明判給侵權的懲罰性賠償，有不同的認定[註10]。

所以在交易實務中，聲明條款是非常重要的。什麼樣的陳述構成聲明，法律課予聲明人如實陳述的義務，若違反義務會有法律上的責任，

---

10 See Dan B. Dobbs, *Law of Remedies-Practitioner Treatise Series*, Vol. 2, §9.2（5）565-68（2d., West 1993）

而什麼樣的陳述則不構成聲明，就算有不實誤導，契約他方不能主張權利？

美國契約法第二次整編（Restatement（Second） of Contract）第159條將「不實陳述」（misrepresentation）定義為不符合事實的宣稱（assertion）。(註11) 由此項定義我們可以推導出所謂的「陳述」（representation）指有關於事實的宣稱。但美國契約法第二次整編第168條又提出「意見（opinion）」的宣稱，在合理的情況之下，可以認定這種意見的宣稱不悖於陳述者所知道的事實，而且陳述者知道的事實足以讓陳述者形成那樣的意見。(註12) 雖然在某些情況下，意見的陳述在法律上也可以合理信賴，但畢竟意見比事實更難繩之以法。

我們可以看到，在前述案例一，法院認為事實的聲明才有拘束力，隨意預估可以養兩千隻羊只是一種意見，意見表述是沒有拘束力的。但在前述案例二，法院卻又認為預估數值這樣的意見，在某些情況下是有拘束力的。

此外不同法域的法律，針對不實聲明，課予聲明人的法律責任可能都不一樣，可能是契約不成立、撤銷契約、解除契約、聲請法院修改契約，或負擔損害賠償責任。因此，締約當事人可能認為法律規定的效果不適合，而想要設計不同於法律規定的不實聲明效果。或者認為不實聲明的區分的類型太瑣細麻煩，不想糾結在故意不實聲明、過失不實聲明、無過失不實聲明，事實陳述、意見表示，隱匿等極其複雜的區別。

---

11 "Misrepresentation Defined. A misrepresentation is an assertion that is not in accord with the facts." Restatement（Second） of Contracts § 159（1981）.

12 "Reliance on Assertions of Opinion. An assertion is one of opinion if it expresses only a belief, without certainty, as to the existence of a fact or expresses only a judgment as to quality, value, authenticity, or similar matters. If it is reasonable to do so, the recipient of an assertion of a person's opinion as to facts not disclosed and not otherwise known to the recipient may properly interpret it as an assertion that the facts known to that person are not incompatible with his opinion, or that he knows facts sufficient to justify him in forming it." Restatement (Second) of Contracts § 168 (1981).

所以契約中的聲明句會以更簡單一些的方式界定來不實聲明定義與效果。

2. 聲明句的結構

　　聲明條款的主要構句要素就是聲明句。在美國契約撰擬實務上聲明句已經發展出相當程度定型化的句型結構：

　　〔聲明主體〕〔依某自然人的認知內容〕〔聲明〕〔聲明內容〕〔聲明有效的特定日期〕〔聲明不實的效果〕

　　以下提供例句一給讀者參考使用，那是作者在美國修課時試擬的合約條款，設定的背景是某個樂團與酒吧老闆商量在酒吧駐唱的交易條件，酒吧老闆要求樂團跟樂手聲明非屬美國音樂家協會會員，因為如果具樂團或樂手會員身份，可能需要按照協會訂定的標準條件來簽約。

例句一：

10. The American Federation of Musicians.

(1) The Band Partnership represents and warrants that, to the Knowledge of the Band Partnership, the Band and all Musicians are not members of the American Federation of Musicians on the Execution Date, Effective Date, and Closing Date of this Agreement.

(2) For the purposes of representation and warranty stipulated in this Article 10 hereof, the "Knowledge" means

(a) General Partner's actual knowledge; and

> (b) The knowledge that General Partner would have had after a diligent investigation.
>
> (3) If afore-mentioned representation and warranty are in any aspects false, incomplete or misleading, the Band Partnership shall pay the Bar USD 10,000 as liquidated damage, and the Bar may rescind this Agreement.

(1) 聲明的主體與自然人的認知內容

前述句型結構中的主體，指的就是契約的一方當事人，這個當事人可能是法人（例如公司），也可能是自然人。如果是法人，因為聲明不實在嚴重的情況下可能構成刑事詐欺。如果是法人的話沒辦法抓去關，公司說假話，就是賠錢了事，沒辦法把公司關進牢裡。所以如果聲明人是法人，而簽約的人認為聲明內容非常重要，就可能要求要具體說明是依據哪一個活生生的人的認知來做聲明，當然這個活生生的人通常必須是代表法人簽約的人，否則契約莫名其妙地拉一個人來聲明，聲明人又沒簽字，發生糾紛時可能就翻臉不認帳了。

而自然人的認知內容也建議需要更細膩的定義。人知道或不知道某件事情，可能有層次的差異。有的是確實知道，有的是大概知道，如果只講確實知道，後來發現情況完全不是聲明的那樣，聲明人可能藉口就他真正知道的，以為是聲明的內容那樣，並不知道不是後來才發現的真實狀況。受害者要證明聲明人在聲明當時腦中實際的認知內容是很困難的，甚至根本是不可能的。所以為了避免聲明人事後找藉口、找理由，通常會將認知更精細地定義為包含實際所知，與審慎調查後應該知悉的內容。

(2) 聲明動作

聲明的動作，中文使用「聲明」，不要含含混混地寫甲方說，乙方說，甲方表示、乙方表示。在英文就用「represent」這樣的動詞，而不是「say」或「state」。

(3) 聲明內容

我們接著討論聲明的內容。關鍵要求是「詳盡、客觀具體的已發生事實，但要適度保留彈性」。

(a) 詳盡

不同的交易需要的聲明內容即有不同，但要完整周延，一般可以考量三個面向（註13），跟當事人有關的聲明、跟契約主要標的有關的聲明，以及跟整個契約能不能履行有關的聲明。跟當事人有關的聲明，例如簽約的公司是否有效登記，合法存續。跟契約主要標的有關的聲明，例如：買賣土地，土地下面沒有隱藏沒清乾淨的污染。跟整個契約能否履行有關的事項，例如：公司跟公司簽約，對方實際出面簽約的人是否獲得有效授權？如果代表簽約人的身份不是很明確，最好是要求出示公司的授權書，簽約的事項依據相關法規、或依據對方公司的內部規章，需要經過何種決策程序，是必須股東會通過、還是董事會通過、或者經理人就可以決定的？需不需要特別的政府許可，例如受到輸出管制？（註14）可以從這種分類方法仔細推敲必須要求聲明的內容。

(b) 客觀具體

---

13 Charles M. Fox, *Working with Contracts: What Law School Doesn't Teach You* 11-13 (PLI Press, 2002).11-13 頁。
14 參閱Fox 著作第171-177頁。

　　如前所述，事實的聲明與意見的聲明在效力讓可能有強弱之別，我們必須盡可能要求事實的聲明，而不只是意見的聲明。我們在路上有時會看到賣水果的、賣蜂蜜的，掛個牌子寫「不甜砍頭」、「不純砍頭」。姑且不論到底要砍誰的頭，但是「甜」跟「純」是什麼？有的人珍珠奶茶微糖（三分糖）就夠了，有人一定要喝到全糖才覺得滿足。去超商買果汁，什麼才叫純果汁？100％不加一滴水，還是90％、70％、50％？賣東西的人常常誇口「品質好」。買東西的人常常說自己「信用可靠」。但發生糾紛時，很難用這種模稜兩可的宣稱主張權利。甜不甜可能是主觀意見，但糖度多少就是客觀具體（註15）。果汁純不純，是主觀意見，但果汁含量百分比就客觀具體。品質好不好，是堅若磐石還是以卵擊石，人言言殊。但有沒有ISO品質認證那張紙，就很客觀具體。「信用可靠」也是隨便說說、隨便聽聽，但專業人士做盡職查核（due diligence），就知道要調閱聯徵中心信用報告，或者查問有沒有惠譽信評、標普信評，或是調票據交換所的票查記錄，甚至是往來銀行出具文書說明有無異常還款。

(c)　已發生的事實

　　而所謂的事實，原則上指的是過去或現在已經發生的事實。未來的事情沒辦法聲明。美國契約法第二次整編第一百五十九條將不實聲明定義為與事實不符的宣稱。而評論C提到，此處的事實包括過去與現在的事實，但不包括未來的事件（註16）。有關未來事情的宣稱，通常只是意見表示或推測，所以聽到這種宣稱的人沒辦法正當地信賴對方的宣稱。要主張權利，必須聽到某項宣稱的人確實信賴宣稱的內容，而且其信賴

15　糖度一般是用20℃下，每100克水溶解的蔗糖克數來計算。https://www.newsmarket.com.tw/blog/65025/，最後點閱日：May 6，2023。

16　*Restatement (Second) of Contracts* § 159 cmt. C (1981).

具有正當性。^(註17)在前述案例二，埃索石油的專家宣稱某個加油站預估銷售量可達20萬加侖。這是未來之事，法院將這解釋為擔保，而課予法律上的效果。

也有學者提出兩種例外情況，一種是關於品質的宣稱，在交易當時，賣方宣稱其產品確實具備某種品質，不論後來賣方做了什麼，或用何種方式交付商品，無論未來使用者如何使用該產品，或發生任何事情，產品的品質都不受影響，則此種聲明就是宣稱未來的事情^(註18)。但其實關於品質的宣稱歸屬於擔保句較為適當。

另一種例外情況則涉及「允諾性的詐欺」（promissory fraud），也就是一方當事人明明知道自己將來不會做某件事，卻允諾會去做。美國契約法第二次整編就明確地將意向納為不實聲明的可能內容^(註19)。有點像是吃霸王餐，一般人走進餐廳點餐，如果沒有意外，吃完就會埋單。但有人存心白吃白喝，也跟一般人一樣走進餐廳點餐，但最後卻用各種手段不付帳。也像是所謂的貿易詐欺，訂貨時就存心詐欺，貨到人就不見了，雖說人心隔肚皮，不是那麼容易證明，但有時候騙得太過明顯。例如：開貿易公司，但辦公室只租一個月，就可以推論詐欺的意圖。當然，是否存心詐騙，還是真有交易上的糾紛，所以不付款，還是要依照個案情況判斷。美國法院通常不願意基於這種理由判賠，因為可能只是真的情況發生變化而違約，卻扯進侵權糾紛，原告甚至還要求懲罰性損害賠償。而在臺灣，民眾也可能用告發詐欺的手段，以刑逼民，其實背後只是違約問題。所以對於這種情況，臺灣的檢察官也不一定會起訴，法院也不一定會判詐欺。

---

17 例如可參閱 *Next Cent. Commun. Corp. v. Ellis*, 171 F. Supp. 2d 1374, 1379-1380 (N.D. Ga. 2001) 轉引自 Stark 著作第 116 頁。

18 *Glen Holly Entertainment, Inc. v. Tektronix, Inc.*, 100 F. Supp. 2d 1086, 1093 (C.D. Cal. 1999) 轉引自 Stark 前揭著第 116 頁。

19 *Restatement (Second) of Torts* § 529,544 (1965)

因此，除非存心未來想要用promissory fraud要求侵權法上的懲罰性損害賠償，或者以刑逼民（當然法院或檢察官不一定願意這麼做），否則關於未來的事，盡量用給付義務句的方式或者擔保句的方式表現較為直接了當[註20]。

(d)　保留彈性

對聲明人來說，當然不宜把話說得太滿。現實世界不是「年年有今日，歲歲如今朝」。聲明人要給自己留點退路。例如：簽授信契約時，聲明列出借款人目前的子公司。萬一哪天關掉或新開一家子公司，那不是借不到錢了[註21]？或是為了要借錢，虧錢的子公司不能收掉，或者拓展事業的計劃必須暫緩？當然事情不能這樣做，需要預先保留彈性。可以把被要求必須聲明，但預期會變動的事項列到附件。然後在本文中說明，這些附件可以隨時由借款人單方更新，或由借款人通知放款人，獲得許可後更新。

有時候聲明的本文或附件，不可避免的必須使用不是最新的數據資料。例如：公司的資產負債表，可能是半年前的。還沒到做新表的時間，也沒辦法逼他做一個出來。這個時候對方可能就要求聲明人聲明資產負債表的事實沒有「重大不利益變動」（material adverse change）。要求簽約時聲明半年前的表在今天仍然屬實，甚至還延伸到履約交割日或其他日期，絕對是切不實際的。所以只能用無重大不利益變動的聲明來替代。「無重大不利益變動」有幾種講法[註22]。

第一種是著重在未發生特殊變化，「自某年某月某日以後，未發生特殊情事單獨地或集合地造成某種狀況的重大不利益變動」（Since

---

20　參閱 Stark 著作第 116-117 頁。

21　參閱 Fox 著作第 14-15 頁。

22　參閱註 3 引用 Adams 著作第 156-179 頁。

[Date], no events or circumstances have occurred that constitute, individually or in the aggregate, a material adverse change in [Field of Change]）。

Field of Change指的是發生變動的領域，通常可以選擇「the business」、「results of operations」、「assets」、「liabilities」、「financial conditions」，也可以更具體化。

第二種是著重在聲明內容準確，例如「有關某事項的聲明並無任何不正確情況，但按照合理預期不會造成重大不利益變動的，不在此限」（Something contains no inaccuracies except for inaccuracies that would not reasonably be expected to result a material adverse change）。

前述無重大不利益變動的聲明也可以改寫做成契約的生效條件。但是，條件不成就，只是契約不能生效，原則上沒有誰賠償誰的問題。但聲明不實，就會有責任。

如果把「重大不利益變動」用契約生效條件的方式來寫，要避免不當地局限了變動的原因。以企業併購來說，買方今天很想把標的公司買下來，但明天可能會有很多原因讓買方不想買了。這些原因可能是標的公司內部的變化、可能是市場情況的變化、可能甚至是買方公司自己內部的變化。

此外，「重大不利益變動」也可能放在給付義務句裡，要求當事人一方，一旦發生重大不利益變動必須通知對方。

也有人用「已產生重大不利益效果的變化、事件或情勢」（change, event, or condition that has resulted in a material adverse effect）替代「重大不利益變動」。

　　重大不利益變動的概念運用與用字遣詞可能相當複雜，包括到底該使用「will」、「would」還是「could」，以及重大不利益變動或效果到底是要寫成已經發生，還是可能發生，以及變動的領域如何界定、變動的效果是個別計算、還是各種雞毛蒜皮的小事加總起來可以作為翻桌走人的理由、重大程度能否量化、該不該量化等等問題，可以從許多訴訟糾紛的判決結果去推敲法院的意向。不過，買方不宜把material adverse change當作萬靈丹，如果想要保留不履約的空間，就想辦法把具體原因預先列出來，沒列出來的事項，法院可能會認為買方也認為非屬重大，所以沒列出來，用「重大不利益變動」、「重大不利益效果」也救不了 ^(註23)。

(4) 聲明有效的特定日期

　　唐朝崔護《題都城南莊》詩是這麼寫的，「去年今日此門中，人面桃花相映紅，人面只今何處去，桃花依舊笑春風」。^(註24) 縱使去年此時聲明這裡有美人，一年後的今天看不到美人，也不能怪聲明人。所以聲明的製造日期與有效期限是很重要的。在美國留學時，作者到China Town的超市買東西，一定要看商品的有效期限，因為遇過慘痛的教訓，買了四罐罐頭，回家一看，有三罐過期很久很久了。簽契約時，同樣也要想清楚，聲明的製造日期與有效期限。一般來說，聲明的製造日

23　*Gordon v. Dolin,* 434 N.E. 2d 341,348-49 (ILL. APP. Ct. 1982)。轉引自註3 Adams著作第178-179頁。在這個案件裡買方在契約裡列了一個生效條件，必須沒有任何標的公司的客戶「實際通知」買方，說客戶不想繼續跟標的公司訂貨，那買方才有義務履行併購契約。併購契約裡也有重大不利益變動條款。買方公司後來從側面管道得知標的公司的大客戶要轉單，就主張構成重大不利益變動，不併了。法院說不行，因為特別約定優先於一般約定。既然退場條件的特別約定說要「實際通知」，這就優先於一般的「重大不利益變動」條款，所以側面管道獲得消息是不算數的。

24　教育部重編國語辭典修訂本，「人面桃花」條目，http://dict.idioms.moe.edu.tw/mandarin/fulu/dict/cyd/0/cyd00535.htm，最後點閱日：April 7，2011。

期就是簽約時，但是，有些契約簽訂之後還會有履約交割日（closing date）。必須寫明履約交割日當天，聲明內容必須仍然屬實。或者替履約交割義務加上前提條件，也就是聲明內容在履約交割日仍屬正確，（註25）也就是把聲明人的聲明送回原廠檢修。在聲明句裡面直接寫清楚，這個聲明句是在多個日期有效（像是簽約日、交割日、與生效日），就像例句一所寫的「on the Execution Date, Effective Date, and Closing Date of this Agreement」。這種做法在美國，稱為「bring down」，有時候除了bring down到履約交割日或生效日之外，也會要求bring down到其他日期，例如貸款契約，聲明日期要包括實際每借一筆款項時。

沒有必要在聲明前面加上「currently」或「presently」。否則，聲明人可能辯稱他的聲明只有簽約日有效，履約交割日就過期了。

也有人用survival條款，想讓聲明內容在履約交割日之後持續保鮮，就是在契約中約定聲明的有效日期延展到履約交割日的某段期間之後（這跟寫出多個聲明做成日期是不太相同的）。但這就做得太過，跟聲明的本質有衝突矛盾。（註26）如果需要在履約交割日之後維持一定的狀況，就直接用給付義務句，課予聲明人維持一定狀況的義務。

由於聲明的事項可能延伸到簽約日之後，在契約實務中為了避免對方事後爭執未來的事不能聲明，所以在聲明條款中常常寫成「聲明並擔保」（represents and warrants）就不會掛一漏萬。但在某些契約中，仍會有只提擔保而不提聲明的事項，通常是用在瑕疵擔保與品質保固（本章第三節說明）。

25　參閱Stark著作第115頁。
26　參閱Fox著作第13-15頁。

(5) 聲明不實的效果

作者獨創太極圖式的理解方法。以「天地陰陽」區分七種基本句型，天指將來的事，地指已發生的事，陰指柔性安排，約定即生效果，無違反可能性。陽指剛性安排，如果違反，即有違約責任。在這樣的理解方式下，聲明句要處理的是已發生的事，而且是剛性的安排。

「聲明不實的效果」至關重要。「聲明不實的效果」這句有兩個要素，「聲明不實」與「效果」。若不約定清楚，對方就會爭執「是否不實」，以及「效果為何」。

什麼是不實？什麼是假話？什麼是說謊？我們可以從政治人物的言行學習不實聲明的藝術。直接做出一個跟已知事實相反的陳述，當然是最差的手段。有些人會隱匿相關資訊，前提條件、以不穩固的前提基礎聲明，顧左右而言他。所以為了應對各種說謊的技巧，在聲明內容中必須鉅細靡遺地列出聲明事項清單，而且在不實效果這部分，需要周延定義什麼是不實？簡單講，就是例句一裡的「在任何方面有虛假、不完整或誤導」（in any aspects false, incomplete or misleading）。雖然律師常常被批評喜歡用同義詞堆疊，擴增文字篇幅以便收取更多費用。但在不實聲明的定義上，「虛假、不完整或誤導」這種三個詞語的串連仍屬必要，因為這三個詞語還是有些細微的差異，而說謊專家很清楚怎麼操作。

確定聲明不實之後，若直接依照管轄法院與準據法來決定效果，那會像賭場裡的輪盤，球到底會滾到哪個號碼的位置很難猜。所以有經驗的律師必然會在契約中約定清楚聲明不實可否撤銷、解除或終止契約（本書之後會再說明撤回、撤銷、解除、終止的意義與差別），能否請求賠償等等事項。

3. 分散的聲明與集中的聲明

如果是交易內容較簡略的契約，聲明事項不多，可能會想到一件說一件，就像例句一裡樂團聲明整個樂團與每個樂手都不是美國音樂家協會會員，這是分散的聲明。而如果是複雜的交易，要聲明的事情很多，就可能集中在契約某一條，完整列出，這就是集中的聲明，通常聲明的事項包括當事人依法設立、有效維持登記，進行的交易是法律允許的，如果需要政府機關特別許可，也已取得相關許可，如果有必要，也經過公司董事會或股東會的決議核准。

4. 主動或被動

寫英文契約有一個大原則，盡量用主動句，不要用被動句。主動句字少、容易懂，比較不會忘記交代做動作的主人。寫聲明擔保句，也是如此。但是有時候，看重的不是誰做的，而是某件事已經被做了。用具體例子來說明，買二手車會看里程，買方不會管是王小明開那台車開了三千公里，還是李大同開那台車開了三千公里，或者他們輪流開，各開了一千五百公里。重點是那台車從出廠之後，只開了三千公里。這個時候，當然就要用被動式來交代聲明與擔保。不然買方反而會被唬弄了。（註27）

5. 句子愈短愈好，愈簡單明瞭愈好

這跟法庭上質詢證人很像。如果一句話寫了十行才斷句，用字艱

27　參閱Stark著作第113-118頁。

澀、賣弄法律術語。[註28] 聲明人就容易找藉口，說是契約文字太拗口，不是故意不實聲明，而且這樣的語句拿給法官看，可能連法官都看不懂，連聲明人到底聲明什麼都講不清楚了，又怎麼能追究欺騙的責任呢？

6. 魔鬼藏在細節裡

　　實務交易常常需要不厭其煩地確認短期、細小的事實，而不要籠統地聲明長期、整體的事實。[註29] 當事人一方聲明公司過去十年是獲利的狀況，可能只有第一年賺一千萬，第二年到最近一年，每年虧一百一十一萬。或者不是專門做地產交易的公司第一年到第九年虧了五千萬，但去年賣掉土地廠房賺了五千一百萬。獲利的具體數額，每一年的營利狀況、本業賺錢還是業外無法持續的一時獲利等，這些可能就是藏在細節裡的魔鬼。

　　在契約上寫聲明條款，有多瑣細的事項，如果疏忽，哪一天就成了法庭攻防的致命傷。許多細節仍然要靠經驗與專業領域知識的累積才會知道。

28　Steven Lubet, *Modern Trial Advocacy* 72-73 (4th ed. NITA 2009)
29　參閱 Lubet 前揭著第 66-68 頁。

第三節　擔保句

案例三^{（註30）}

推土機瑕疵擔保案

　　買方向賣方買了一台一般規格履帶式推土機。交機之後，賣方出具保固書，上面記載「貴局購置履帶式推土機壹台，本公司保證自驗收合格日起1年內，在正常操作使用情形下對推土機本身之結構、零組件負責保固1年，在保固期間如有故障，本公司負責免費修復或更換新品。但故障或損壞如係因貴局人為操作不當或未依原廠規格進行維修所致、或未使用原廠零配件致推土機功能減損者，本公司將不負保固責任。」並依照契約約定繳付保固保證金5萬9500元給買方。

　　買方驗收後約兩個月餘推土機即陸續發生冷卻液洩漏、水溫過高及控制桿故障，賣方修好了，也同意延長保固18日。但之後約再3個月至7個月間又發生引擎冷卻散熱器冷卻液及操作油溫度過高、推土機無法前進後退等故障，買方限期要求賣方修復，賣方未修復，買方依民法第226條第1項、第256條等規定主張給付不能而解除買賣契約，並請求返還價金並支付解除前的遲延交付利息。

　　賣方的抗辯是保固書只是附約，所以要主張給付不能解除契約必須依照主約的約定。推土機已經交付並驗收通過，買方事後要主張推土機有瑕疵必須證明那樣的瑕疵在交付的時候就已經存在，也要證明推土機完全無法使用，並證明瑕疵的發生不是因為

30　臺灣高等法院97年度上字第447號民事判決，及最高法院98年度台上字第1214號民事判決。

買方操作不當。賣方也主張因為買方保管不當，推土機才嚴重毀損難以修復，包括買方沒有用防水帆布或塑膠套防護，讓推土機任憑風吹日晒雨淋，造成推土機嚴重鏽蝕不堪使用、無法修復。並非可歸責於賣方致給付不能。就算賣方給付不能，買方對於損害的發生與擴大也與有過失。

　　法院首先認定契約中的義務分成三類，主給付義務，從給付義務與附隨給付義務。主給付義務不履行，可以解除契約，附隨給付義務不履行只能請求損害賠償，不能解除契約。其次基於保固書的文字只說對推土機本身結構、零組件負保固責任，保固期間內如有故障，賣方負責免費修復或更換新品。而故障時，買方發通知給賣方也指出如果賣方未善盡履行保固責任，將逕行動用保固保證金另請第三人修復，以此推論買方主觀認知賣方的責任在於修復。所以法院認為本件賣方的保固責任屬於附隨給付義務。

　　至於第一次修復完成後，又再次故障，雖故障具體原因不明，但疑屬推土機本身設計有瑕疵。雖然賣方主張買方必須證明，但法院認為應該是賣方要負責證明買方操作不當。所以買方主張解除契約、退款並請求遲延利息為無理由。只能從保固保證金5萬9500元取償。

1. 擔保的目的與法律效果

　　以上案例顯示擔保條款的重要性，擔保條款將責任範圍寫得大，責任就大；擔保條款將責任範圍寫得小，責任就小。而擔保條款的構句要素就是擔保句。我們首先必須釐清中文「擔保」與「保證」這兩個詞，它們口語上常常混淆，甚至連法律文字都是混淆不清的。本書所指的擔

保是指一方當事人向他方具體確保未來會發生或不會發生某種情況，而這種情況不是一方當事人要做什麼事或不做什麼事，不是擔保的那一方當事人可以掌控的事。因為如果是可以掌控的事，是作為或不作為，那是另一種句型，稱為給付義務（covenants）。但因為無法掌控，所以如果事情發展不如擔保的那樣，擔保的人就有其他責任。而常見的擔保是提供商品、權利、服務或資訊的一方，向他方具體確保提供的商品或服務沒有瑕疵，或者有特定的品質，或者能持續一段時間維持具體指定的功能或功效，如果有瑕疵或不符合約定的品質，或不能維持具體指定的功能或功效，則提供商品、權利、服務、資訊的一方必須負責修復、更換、或退款。而保證指的則是一人承諾為他方的債務負某種責任，例如：債務人不還錢的時候，保證人負責還錢。如果完全沒有白紙黑字的擔保條款呢？聲明在法律上可能有通則性的規範，例如：臺灣民法第92規定「1.因被詐欺或被脅迫而為意思表示者，表意人得撤銷其意思表示。但詐欺係由第三人所為者，以相對人明知其事實或可得而知者為限，始得撤銷之。2.被詐欺而為之意思表示，其撤銷不得以之對抗善意第三人。」不論在什麼樣的交易中被詐欺而為意思表示，都適用這樣的規則。但擔保的法律規範可能在不同的交易類型中有不同的規範，例如：臺灣民法第354條針對買賣這樣的交易規定：「1.物之出賣人對於買受人，應擔保其物依第373條之規定危險移轉於買受人時無滅失或減少其價值之瑕疵，亦無滅失或減少其通常效用或契約預定效用之瑕疵。但減少之程度，無關重要者，不得視為瑕疵。2. 出賣人並應擔保其物於危險移轉時，具有其所保證之品質。」臺灣民法有時用「擔保」這個詞，有時是用「保證」這個詞。例如前述第354條第2項，所以用語並不一致。如果賣方擔保內容不實，買方可以解除契約，或者請求減少價金，或者請求不履行之損害賠償。^(註31)

31　民法第359條、360條。

　　美國統一商法典§2-313（註32）則對於什麼是明示的擔保做出規範，賣方針對某個事實向買方做的任何確切的表示或允諾，而成為交易的基礎，就構成明示擔保，所以銷售的產品必須符合該項表示或允諾。如果對產品做任何描述，而這樣的描述成為交易的基礎，則這也構成明示擔保，銷售的產品必須符合該項描述。如果用樣品或模型作為交易的基礎，同樣也構成明示擔保，銷售的產品必須符合提供的樣品或模型。

　　要構成明示擔保，賣方並不需要特別使用「擔保」（warrant）、「保證」（guarantee），或其他特定的字彙。但如果賣方只是對於產品的價值做確切的表示，或出於自己的意見或賣瓜者言而說一些話，就不構成擔保。如果賣方違反擔保約定，在具體情況下負擔的責任可能包括換貨、修理、退款、減價等等。（註33）

2. 擔保句的結構

　　擔保句的造句方式及寫作訣竅大致上可以跟聲明句相似，也就是：〔擔保主體〕〔擔保對象〕〔擔保〕〔擔保內容〕〔擔保時點或期間〕〔違反擔保的效果〕。如以下例句二所示：

---

32 (1) Express warranties by the sellers are created as follows:
　　(a) Any affirmation of fact or promise made by the seller to the buyer which relates to the goods and becomes part of the basis of the bargain creates an express warranty that the goods shall conform to the affirmation or promise.
　　(b) Any description of the goods which is made part of the basis of the bargain creates an express warranty that the goods shall conform to the description.
　　(c) Any sample or model which is made part of the basis of the bargain creates an express warranty that the whole of the goods shall conform to the sample or model.
　　(2) It is not necessary to the creation of an express warranty that the seller use formal words such as "warrant" or "guarantee" or that he have a specific intention to make a warranty, but an affirmation merely of the value of the goods or a statement purporting to be merely the seller's opinion or commendation of the goods does not create a warranty.
33　美國統一商法典§§2-316, 2-718, 2-719.

例句二：

Warranty. The Manufacturer warrants to the Retailer [and the eventual consumer] that the Products as packages and shipped from the Manufacturer's plant *will be free* from substantial defects in material, design and workmanship and *will function and perform* in accordance with Manufacturer's specifications for a period of one year from the date of retail purchase（hereinafter, the "Warranty Period"）under usage in manners compliant with user's manual. If within the Warranty Period, the Products are with any substantial defects in material, design or workmanship or the Products may not perform in accordance with Manufacturer's specifications and (1) if such defects or nonperformance are informed by the Retail within 1 month after the date of retail purchase, the Manufacturer shall replace the defective Products or refund the Price at the Manufacturer's discretion, or (2) if such defects or nonper-formance are informed by the Retailer to the Manufacturer after expiration of one month after the date of retail purchase, but prior to expiration of the Warranty Period, the Manufacturer shall repair, replace the defective Products or refund the Price at the Manufacturer's discretion.（註34）

---

34 參閱註Stark著作第118頁並調整修改。

(1)擔保主體

擔保主體當然是提供產品、資訊、勞務或權利的一方。

(2)擔保對象

聲明句的聲明對象理所當然就是簽約的當事人，但擔保有時候可能不只是簽約當事人，也可以約定擴及最終的買家。

(3)擔保的動作

在中文使用「擔保」，在英文就用「warrant」。

(4)擔保內容

需留意擔保的內容可以分成兩大類。一大類是瑕疵擔保，擔保在交付時，交付的標的沒有各種瑕疵，而所謂的各種瑕疵則可再分成物之瑕疵與權利瑕疵兩類，物之瑕疵通常會約定成(1)製造上的瑕疵、(2)設計上的瑕疵、與(3)材料上的瑕疵三種，而權利瑕疵則指(1)賣方對標的物欠缺完整正當的所有權與處分權,(2)標的物上有抵押、質權、留置等擔保物權，以及(3)第三人對買方主張標的物侵害其專利權、商標權、或著作權等各種智慧財產權，而這樣的擔保條款通常又要搭配彌償條款（Indemnity），亦即規定如果有第三人向買方主張侵害各種權利（所有權、擔保權、智財權），則賣方應該補償買方因此所支出的一切成本與費用及各項隨之而來的損害。擔保的另一大類則是品質保固的擔保，那是指在交付後的約定期間內，交付的標的繼續符合約定的品質、規格與效能，也就是通常所說的保固。以下以圖示說明以利理解：

一、瑕疵擔保
  1.物之瑕疵

    (1)製造瑕疵
    (2)設計瑕疵
    (3)材料瑕疵

  2.權利瑕疵

    (1)賣方欠缺完整正當的所有權
     與處分權
    (2)標的物上有抵押、質權、留
     置等擔保物權
    (3)第三人對買方主張標的物侵
     害其專利權、商標權、或著
     作權等各種智慧財產權

二、品質保固 交付後的約定期間內，
      交付的標的繼續符合約
      定的品質、規格與效能

圖2 擔保分類圖說

  以前述案例三的事件來說，法院一面認定推土機已交付通過驗收，所以只剩下品質的問題，但也提到可能有設計上的瑕疵，卻未深究是否真的有設計上的瑕疵，而直接認定依照保固書上的修理方式處理，這在法律上還是有問題的，因為設計上的瑕疵，也可能是違反瑕疵擔保而有解除契約的事由。

  不同法域的法律可能對於瑕疵擔保與品質保固的擔保有不同的規定與要求，可能內容很複雜，例如區分明示擔保與默示擔保，而且成立擔保的要件可能相當複雜，例如美國統一商法典規定若賣方是商人，該商人的業務內容就是銷售某類商品，那麼他在銷售那一類商品的時候，除非契約另外約定，就已默示擔保該商人所銷售的商品具有適銷性（merchantability）[註35]。當事人可能希望有簡單明確一些的約定，不

---

35 U.C.C. § 2-314 (1).

管賣方是不是商人，也不論賣方的常態業務是不是銷售這份契約的標的產品，我們都要賣方擔保產品符合一定的品質、規格或功能，而且我們可能不希望用適銷性這種難以理解的用語，而希望更明確約定具體的規格、功效、包裝的方法等等。如果沒約定清楚，往往是發生事情，才找契約準據法條款，再去翻法條，看看法律是怎麼規定的，結果發現完全不符合交易的需求。因此就買賣的擔保內容，當事人可能約定在交付時點，交付的產品沒有物之瑕疵或權利上的瑕疵，而在交付後（或零售購買日後）一段期間內，產能仍能持續符合約定的品質、規格或功能。此處需要特別考慮保固內容的寫法，避免過於曖昧的說法，像是不會故障，雙方對於什麼叫做故障可能有不同的理解，手機觸控螢幕碰觸到反應作用之間時間多久才算故障，所以建議在附件中列出明確的品質、規格、功能之說明，加以參照，較不會發生糾紛。另外，持續符合約定品質、規格或功能必然有相關的條件或例外排除情況，例如人為不當使用，當然不能保固，拿榔頭敲手機，或將手機從高空丟下、泡水，肯定是不應該保固的，不然賣方會賠不完，但是這些例外情況的說明也必須詳盡而明瞭，如果有個使用手冊，當然是最能避免疑義。

(5) 擔保時點或期間

如前所述，擔保通常有兩種內容，瑕疵擔保與品質保固的擔保，瑕疵擔保的時間點是指交付時沒有瑕疵，而交付後的擔保只限於品質保固的擔保。品質保固擔保的期間可能是從交付、零售購買日，註冊登記日，或其他起算點計算一段期間，如果是汽車也可能是按里程數計算，或者期間跟里程數併用，取先到的時點。

(6)違反擔保的效果

不同法域對於擔保的效果有不同的規定，就像前述案例三，法院認定違反擔保只能請求修復或損害賠償，不能解除契約。所以為了避免法律給你的效果不符你的意，你可以在契約中寫清楚，若違反擔保是要解除契約、終止契約、還是修復、賠償、減價退款，就不受法律限制。

本章說明聲明句與擔保句的目的、法律效果與句構方式。但英文契約常見另外一種與聲明、擔保有關，但方向全然相反的語句。也就是免責聲明（Disclaimer）。契約中之所以需要寫免責聲明，那是因為法律規定了默示擔保，有時候交易主體什麼話也沒說，但法律推定他們說了。例如美國統一商法典規定若賣方是商人，該商人的業務內容就是銷售某類商品，那麼他在銷售那一類商品的時候，除非契約另外約定，就已默示擔保該商人所銷售的商品具有適銷性。所以如果真的不擔保默示擔保的內容，就要在契約中特別明白地排除，那種句子就是免責聲明了。

以下提供免責聲明的例句：

例句三：

SELLER DISCLAIMS ANY WARRANTY OF MERCHAN-TABILITY OR WARRANTY FOR FITNESS OF USE FOR A PAR-TICULAR PURPOSE, EXPRESS OR IMPLIED, WITH RESPECT TO [PRODUCT NAME].

為什麼要全部用大寫粗體英文字呢，有時候是因為法律要求要免責聲明必須顯著。[註36] 當然，要讓契約文字顯著不見得就是全部大寫粗

---

36 U.C.C. § 2-316 (2).

體，不過這是很常見的用法。有時候法律真的要求要大寫粗體，像亞利桑那州的法律規定出售住宅的契約如果有約定採用法院訴訟以外的紛爭解決條款，該條款必須全部大寫、粗體，並且要十二號字以上。[註37] 不過也有學者認為全部用大寫，讀者連看都不想看。而且也有法官在實際判決對全部大寫頗不以為然。[註38] 因此該學者主張除非是法律明文要求，否則，用粗體加斜體來表示強調比較好。[註39] 所以我們也可以把免責聲明寫成：

關乎未來的事。例如：

例句四：

Seller disclaims any warranty of merchantability or warranty for fitness of use for a particular purpose, express or implied, with respect to [Product Name]

並不是所有的責任都能用免責聲明排除掉的，在美國，商人賣產品給消費者時，如果在契約中聲明產品有問題造成消費者生命、身體、健康受損害，也不賠償的話，這種免責聲明原則上被當做是悖於良知的，所以不會發生效果。[註40] 另外，如果當事人在契約裡一方面擔保自己的產品有什麼樣的品質，另一方面卻又聲明免責，法院會把這兩種有衝突的聲明擺在一起解讀，免責聲明通常會被剔除當做無效。[註41] 美國統一

37　Ariz. Rev. Stat. § 12-1366 (A)(1).

38　"Lawyers who think their caps lock keys are instant "make conspicuous" buttons are deluded. In determining whether a term is conspicuous, we look at more than formatting. A term that appears in capitals can still be inconspicuous if it is hidden on the back of a contract in small type." **In re Bassett, 285 F.3d 882 (C.A.9. 2002)**

39　參閱註3引Adams著作第350-352頁。

40　U.C.C. § 2-719 (3).

41　U.C.C. § 2-316 (1).

商法典也規定該法所要求的誠信原則、勤奮謹慎、合理性與注意義務是不能聲明免除的。[註42]

免責聲明的功能是要排除默示擔保，但是由句型分類的角度來看，免責聲明既非聲明句，也不是擔保句，而是宣示句。因為免責聲明是柔性的條款，沒有違約的可能性，沒有違反的效果，約定了就有效。在之後的章節我們會再解說宣示句。

最後補充說明，雖然我們將擔保句列為一種獨立的句型，但其實那與給付義務句是非常近似的，擔保句是關乎未來之事，是剛性條款，違反會有違約責任。而給付義務也是關乎未來之事，也是剛性條款，違反會有違約責任，所以如果將擔保句與給付義務句同一種句型也不是不可以，也就是說擔保句是加了條件的給付義務，如果發生特定情況，例如在保固期間內，正常使用，產品不能發揮合意的功能，則擔保人有義務修復、換貨或者退款。而給付義務通常是在特定時間必然有義務為特定行為，或不為特定行為。但擔保句往往採用固定的動詞「擔保」（warrants），而且有相當固定的內容，瑕疵擔保與品質保固擔保，所以在本書中我們將擔保列為獨立的句型。

---

42 U.C.C. § 1-302 (b).

給付義務句

第一節　給付義務句的意義

案例一

Harrison vs. Land's End of Emerald Isle Association, Inc.
（註1）

　　原被告是北卡羅萊納州加特利郡豪華海景社區的三戶住戶，原告起訴指稱兩戶被告違反社區規約第十三條，該條規定如下：

　　「每一塊土地的所有權人應該以乾淨而悅目的方式，維護他的土地與上面的結構物，而且應該除草，所有結構物都應該用能夠維持與其他良好維護的土地，與結構在美觀外表上相容的方式上漆並照管。違反者可能被宣告為「持續擾鄰」直至改正為止」。

　　被告的兩戶人家，一戶在自己的地上密集地種植橡樹、蠟香桃木，還有其他一些植物，但似乎很快地乾枯了。而另一戶則土地有些積水，種滿香蒲草。而且地上設了擋土牆，有時豪大雨之後會從牆底滲入，甚至是積水將近十個月，最後才蒸發或完全滲入地下，因為這樣的情況，原告起訴指摘被告住戶，違反前述社區規約。

　　原審判原告敗訴，原告上訴後。被告主張社區規約模糊不清所以無效，不能強制履行。上訴法院也認同被告主張。

　　上訴法院並引用相關的前案見解「給付義務這個字指的是有利於締約雙方而有拘束力的合意或約定」（*Armstrong v. Ledges Homeowners Ass'n*, 360 N.C. 547, 554, 633 S.E.2d 78, 84 (2006)）"依據契約法一般原則，契約的條款必須充分確定，法院

1　https://sellersayers.com/hoa-law/warning-covenants-too-vague-to-enforce/，最後點閱日：May 30，2023。

才能強制履行"（*Snug Harbor Prop. Owners Assoc. v. Curran*, 55 N.C. App. 199, 203, 284 S.E.2d 752, 755 (1981)）「給付義務的內容如果沒有『清楚而不曖昧』的標準讓法院可以『客觀判斷』當事人是否符合其標準，即屬因為模糊而無效。」（*Beech Mountain Prop. Owners' Assoc. v. Seifart*, 48 N.C. App. 286, 295,269 S.E.2d 178, 182 (1980); *Lake Gaston Estates Prop. Owners Ass'n v. Cty. of Warren*, 186 N.C. App. 606, 652 S.E.2d 671(2007).）

上訴法院認為社區規約要求所有權人維持「乾淨而悅目」的狀況，以及「與其他良好維護的土地與結構在美觀外表上相容」這樣的標準很模糊，可能因為每個人主觀偏好而有不同解釋，所以給付義務中欠缺可確認的標準，讓法院客觀判斷被告行為，所以法院若允許強制履行會流於恣意。上訴法院結論認定本案的給付義務過於模糊而無效。

「一方應為他方做某件事，或不做某件事」，就是一項允諾。在寫契約時，就用給付義務句（covenant）來寫。給付可能是作為，也可能是不作為，不作為其實就是禁止。

給付義務（Covenant）這個字的意思是一種慎重的承諾，承諾作某件事，或者不作某件事，而且源自於蓋印，因為在承諾的文件上蓋了章，所以顯得特別慎重。通常蓋章是用在轉讓不動產產權的契據上。（註2）而且因為有外觀上的慎重性，所以依照英美的普通法，即使沒有約因（對價），蓋了章也是有效可強制履行的契約。（註3）

2　Holmes, Eric M. (1993). "Stature and Status of Promise Under Seal as a Legal Formality". *Willamette Law Review*. **29**: 617.

3　"Developments in the Law of Seal and Consideration in New York". Cornell Law Quarterly. 26: 692. 1941.

　　給付義務句是關乎未來之事,而且是剛性的內容,違反即有法律上的責任。

　　義務與權利是一體兩面,但在撰寫契約時,出發點不同還是會有一些差異,所以既有給付義務句,也有權利句。主要的原因在於給付義務的對立面,是受領給付,而不是權利;而受領給付除了有權利之外,也有義務,若不按時受領,受領遲延,造成給付人費用增加,受領人可能也需要負擔損害賠償責任。但契約中的裁量選擇句,則是處理純然享受權利的情況,例如有終止契約的權利,不會因為不行使而違約需負擔責任。所以,裁量選擇句這樣的權利語句是關乎未來之事,但卻是柔性的,違反並不會有責任。

第二節　給付義務句的結構

例句一:

　　Weekly Music Performances. On Saturday, November 14, 2015, and each following Saturday during the Term, the Band Partnership shall direct Formosa to make two Sets of Music Performances at the Bar from 8:30 pm to 9:30 pm and from 11:00 pm to 12:00 pm respectively without break during each set of Music Performance.

　　例句一就是一個給付義務句。關於給付義務句的組成元素,許多人提出各種主張,有的主張是「event」,「principal」,「act」,「resource」，與「condition」，（註4）也有說是「who」、「what」、「when」、「where」、「why」、「how」，與「how much」。（註5）華人習慣的「人、事、時、地、物、數」也可供參考。

我們可以將上述幾種說法整併起來，並加上缺漏的受領人，應該更為完整，也就是：

- 義務人（who or principal）

- 表示義務的助動詞（why）

- 履行義務所涉及的動作（act）

- 給付標的（resource or what）

- 標的數量（how much）

- 標的之品質

- 受領人（whom）

- 履行義務的地點（where）

- 履行義務的方法或標準（how）

- 履行義務的時間（when or event）

- 義務發生或消滅的條件（condition）

以下我們擇要說明相關元素的考量點。

4　See, Xing Wang et al., *The Contract Expression Language –CEL*, http://www.contentguard.com/ drmwhitepapers/The_CEL.pdf (accessed 2011/3/20)，CEL 是一項研究以電腦寫出並判讀契約的計劃。

5　Tina L. Stark, *Drafting Contracts: How and Why Lawyers Do What They Do* 125 (Wolters Kluwer, 2007).

1. 義務人

　　給付義務的義務人必須是在前言裡交代的締約當事人。用例句一來說明，假設有一個新創立的搖滾樂團用「Formosa」這個名字行走江湖，但是這個團員為了做生意方便，用合夥方式做商業登記，它的英文正式名稱是「Formosa Partnership」。假設後來Formosa Partnership人氣飆升，又組了「Formosa Junior」這個新樂團，找一些有潛力的二軍團員提供不同的表演，我們就會發現「Formosa Partnership」跟「Formosa」樂團可能不是完全的相同。有一家酒吧想要聘請「Formosa」樂團駐唱，在契約裡，我們的締約當事人是合夥事業，簡稱為「Band Partnership」。如果我們在契約裡直接約定Formosa Partnership應該提供什麼樣的表演，Formosa Partnership可能派二軍Formosa Junior充數。如果我們指定要Formosa裡面的各個團員，可能團員又不願意在契約上簽字，只有Formosa Partnership簽字，不見得拘束得了個別團員。（註6）所以就發生實際履行的人，跟契約義務人不同的狀況。

　　重點在於「義務人必須是締約當事人」，所以在英文契約裡我們可以寫「the Band Partnership shall direct Formosa」或「the Band Partnership shall cause Formosa」，以表示締約當事人「Band Partnership」有義務讓第三人「Formosa」做出指定的事。

6　在不同國家，合夥可能有多種組織方式，不見得所有合夥人都負連帶責任。

2. 表示義務的助動詞

英文契約裡，就用「shall」來表示做某件事的義務。[註7]不要用「will」，因為「will」是願意做，但沒有義務的成份。至於「agrees to」、「is responsible for」、「undertakes to」、「covenants to」、「shall be obligated to」或「covenants and agrees to」[註8]也是應該做的意思，只是較為冗贅。「must」是用在條件句，表示必須發生一定的前提才有特定的結果，所以不適合用「must」來替代「shall」。總之，表示義務時，就用「shall」，不要用其他任何詞語。不表示義務時，就不要用「shall」。

「shall」用的對與錯有幾個簡單的判斷準則。[註9]

(1)　「shall」前面只能是締約當事人。

(2)　「shall」後面不能加「have」：「have」有兩種意義。一種是用來表現完成式，義務是未來的義務，所以不能用完成式。如果意思是「必須完成特定條件，契約才生效」，那麼用「have」來提示一種前提條件時，在「have」前面不能用「shall」，而要用「must」，例如「A must have obtained consent of Parent Company」。「have」的第二種意義是用來表示「享有」，在契約裡通常是表示享有特定權能，「have the right to」。此時也不需要用「have」，直接用「may」，例如「A

7　參閱註5引 Stark 著作第126至130頁。也可參閱 Kenneth A. Adams, *A Manual of Style for Contract Drafting* 43-45 (3rd ed., ABA, 2013).

8　也有人主張 "shall" 在文法上有兩層作用，一層表示義務，一層表示未來的時態。在表現未來時態時，第一人稱才用 "shall", 第二人稱、第三人稱使用 "will"。因此，可用 "has a duty to" 來表示義務。此外，也有人為了避免英文契約用語跟一般英語用法脫節過大，所以用 "must" 來替代 "shall"，畢竟 "must" 沒有限第幾人稱使用的問題。參閱註7引 Adams 著作第33-38頁。不過使用 "shall" 表示義務，仍是普遍的共識，如果不表示義務時，就不要用 "shall"，應可避免時態人稱上的困擾。

9　見註5引 Stark 著作第150到153頁。

may terminate this Agreement」，而不需要說「A has the right to terminate this Agreement」。但絕對不能說「A shall has the right to terminate this Agreement」。

(3) 「shall」後面不能加「be」：「be」可能帶出一個被動式，也可能是做出某種定義、或宣示某種政策。定義要用宣示句，沒有義務的成份。常見誤用「shall」做宣示的例子就是準據法或管轄法院條款「Any disputes arising from this Agreement shall be governed by applicable laws and regulations of Taiwan」。可以把「shall be」改成「is to be」，或「will be」。被動式不能成為義務，沒有「賣方應被支付價金」這回事，只有「買方應支付賣方價金」。如果「賣方應被支付價金」，卻沒有被支付價金，是誰違反義務？從字面上來看是賣方違反義務了。當然有一種情況，一方有義務允許另外一方做某些事，我們可以用「A shall allow B to」來表示，無論如何，「shall」之後不能加「be」。

(4) 條件子句裡面不能用「shall」：「shall」只表示未來的義務，條件跟義務是大不相同的，條件不符合常常是「天要下雨、娘要嫁人」，沒辦法的事，最多只是「買賣不成仁義在」，已經給付的部分可能用不當得利要回來，該守的保密義務還是要守。但如果違反了義務，就會產生責任。所以不要誤用「shall」，免得對方有意或無意地誤把條件當義務。

另外，「shall」直接加上動詞，代表應作為的義務。「shall not」加上動詞，代表不作為的義務，也就是禁止做什麼事。契約禁止做什麼事是一回事，當事人會不會做被禁止的事是另一回事。所以才要以違約責任相繩，但有時候違約責任還不夠。例如房東將房子租給房客使用，並約定房客不得將房子轉租或租，也就是房客沒有權利再轉租、分租。但房客即使沒有權利（right），卻還是有實力（power）。房客把房子轉

租、或分租出去，房東怎麼辦？所以除了用給付義務句「shall not」加以禁止之外，也可以用宣示句將轉租、分租效力約定為無效（註10），當然這部分可能必須查詢各國法律，對於善意的交易相對人的保護規定。或者也可以在契約中給予房東終止契約的權利，一旦發現房客當起二房東，直接收回房子。

3. 履行義務涉及的動作

如前所述，給付義務句當中，在「shall」之後，用來表現動作的動詞不能是「be」或「have」。

給付義務句中，主詞、「shall」，動詞、與受詞應盡可能緊密相連，這才是最容易讓人看懂的方式。一方面不要將主詞、「shall」，動詞、與受詞這些元素硬生生地拆散。另方面每個元素不要寫得又臭又長。

履約的相關動作盡可能用動詞來表現，有些人喜歡用名詞、甚至形容詞來表現動作，這是很不恰當的。因為「shall」不能直接加名詞或形容詞，所以要再加字。甚至還得改成被動式，那就更多字了，而最自然的「主詞、「shall」、動詞、受詞」構句就被破壞了。輕則造成腦部運動量增加，嚴重的甚至扭曲了重要的意思。例如，會計記帳要講「Account Payable」或「Account Receivable」，但在契約裡，避免用「-able」給自己找麻煩。「可以付的款項」跟「應付款項」是有差別的。「可償還的費用」、跟「應償還費用」在強度上是不一樣的。尤其在給付義務句裡，切記要避免把動詞埋掉了。

10　參閱註5引 Stark 著作第171頁。

4. 給付標的

　　依照臺灣的法律，法律行為的標的必須確定、適法、可能、妥當。在美國，契約的主要內容大致上也是要符合這些要求。在契約法下，要約的主要內容必須明確。隨著契約交易標的之不同，也有不同的標準以判斷標的是否明確。

　　但是標的概念本身確實太過模糊難以掌握，無論是在實定法律，或在法學院的訓練中，都缺少清楚的給付分類，所以契約中關於給付標的常常寫不到重點，就像本章案例一的情況。

　　臺灣民法債編第一章第二節規範，債之標的第199條規定「1.債權人基於債之關係，得向債務人請求給付。2.給付，不以有財產價格者為限。3.不作為亦得為給付。」所以繞了一圈，標的就是給付，給付就是標的，還是讓人霧煞煞。只是本條第三項多說了一句不作為也可以給付。臺灣民法第200條講物、第201條到第207條講的都是貨幣、利息，簡言之就是錢。然後就沒有任何規定提示給付標的到底是什麼。但從存在的規定我們可以找到幾種標的：錢、物、不作為、與反面的作為。

　　因此筆者思考後提出更為明確周延的給付標的分類法：物、金、資訊、人、權利。物當然就是各種貨物、商品、產品；金就是金錢，也包含虛擬貨幣；資訊就是以各種資訊作為標的；人不是販賣人口，而是指的人的行為，行為通常可分成賣時間（僱傭）、賣做特定的事情（委任）、賣做出特定的成果（承攬）；至於權利則是指法律規定的各種權利，例如物的所有權、使用權、擔保權，智財權（著作、專利、商標、營業秘密）。上述五種標的之分類可以是做正面的約定，例如應交付物、應交付金錢、應交付或拋棄權利、應交付資訊、應做特定行為。但也可以做反面的約定，例如不應處分物，不應移轉資金、不應洩漏資

訊、不應處分權利、不得競業。

在個人觀察過的契約中，給付義務內容可說，不脫物、金、資訊、人、權利這五種標的之作為與不作為，所以這樣的分類相當實用。

5. 標的之數量、計算公式與品質

就土地交易來說，土地跟價金都必須由當事人明確地加以指定。就貨物銷售來說，在締約時，數量必須確定，或者至少有方法可以確定。有一類交易較為特殊，涉及需求契約（requirement contract），或產出契約（output contract）。需求契約指的是由某家供應商完全供給某個客戶的需求，產出契約指的則是由某家客戶完全買下某個供應商的全部產出。在這類契約中，即使並未明確提到標的物數量，但還是可以參考客觀、外在的事實加以特定，例如供應商過去實際的產量，或者客戶過去實際的需求量。在這類契約中，雙方都依照誠信原則進行交易，供應商不能趕快擴產、拼命交貨。客戶也不能突然減縮訂單。由於誠信與默契，在這類契約占有很大的重要性，所以某些法院不會執行新成立的企業，所訂定的需求契約、或產出契約，因為新企業的產出實績或實際需求數量，沒辦法依照過去的數據加以推估。

對於需求契約或產出契約，固然在美國的契約法之下，法院可以參照過去的客觀外在數據加以判斷。但是當事人自己訂出一個參照值，當然是更妥當的。

契約中，算數的問題非常複雜，會因為不同交易類型而有變化，而且也適用於契約各種基本句型，而不只與給付義務相關，因此在其他章節會再詳細說明。

至於標的之品質，通常是技術性的約定，不同的標的會有不同的約定。如果是專業的買家自然會約定妥當，但是通常買方對標的相關資訊的掌握程度是不如賣方的，買方該如何處理呢？一種方法是在締約背景中特別說明標的指定用途，擔保條款主要是可以從法律有關默示擔保的規定部分加以推敲，細節我們在後續相關章節再說明。

6. 受領人

如果受領人就是締約當事人，這是最正常的情況，因此不需要特別著墨。但是在交易中，給付對象時常不是締約的當事人，而是第三人。比如說甲男買花，請花店老闆乙，直接把花送交甲的妻子丙。

雖然有在臺灣法律明文規定（註11），以契約訂定向第三人給付，第三人可以直接請求給付，但是法院實務卻又區分「利益第三人契約」的情況，與「指示給付關係」，前者的第三人可以直接請求交付，後者的第三人不能，而判斷的標準在於考量契約訂定之本旨，是否由第三人自己行使權利，比僅由要約人行使權利，更能符合契約之目的。（註12）美國

---

11 臺灣民法第269條規定：「1.以契約訂定向第三人為給付者，要約人得請求債務人向第三人為給付，其第三人對於債務人，亦有直接請求給付之權。2.第三人對於前項契約，未表示享受其利益之意思前，當事人得變更其契約或撤銷之。3.第三人對於當事人之一方表示不欲享受其契約之利益者，視為自始未取得其權利。」

12 參閱最高法院97年台上字第2694號判決。「審認契約是否有以使第三人取得該債權為標的，並不以明示為必要，祇要依契約之目的及周圍之情況，可推斷當事人有此法效之意思為已足。於此情形，除審究其契約是否為第三人利益而訂立外，尚可考量契約訂定之本旨，是否由第三人自己行使權利，較諸僅由要約人行使權利，更能符合契約之目的，債務人對第三人為給付是否基於要約人亦負擔相當之給付原因暨要約人與第三人間之關係，並分就具體事件，斟酌各契約內容、一般客觀事實、工商慣例、社會通念等相關因素，探究訂約意旨之所在及契約目的是否適合於使第三人取得權利，以決定之。」

契約法第二次整編也列了相似的規則。^(註13)因此關於第三人可不可以直接請求，以及債權人能不能變更受益人，在具體個案中，不同法域可能有不同的判斷標準，而且那些標準常常是一言難盡、非常複雜。所以為了避免在發生糾紛後才由法官選法作出超乎當事人預期的決定，不如當事人自己在契約中寫清楚，第三人可以或不可以直接向債務人請求，以及債權人可不可以變更受益人，什麼時點前可以變更，什麼時點後不能變更。

　　還有一點需要提醒，在商人對商人之間的買賣，交貨給第三人通常不會有太大問題，但是約定付錢給第三人，可能就要想想稅務機關會不會認為意圖逃稅，或甚至洗錢。

7. 履行地

　　履行地也是很重要的，重要性不僅在於實際操作上，在哪裡履行會影響成本，將貨運送到臺灣本島跟運送到離島當然是不同的，但除此之外，履行地也會影響風險的移轉、所有權的移轉，甚至是準據法或管轄地。縱使雙方約定將貨最終送到特定地點，但每一段的運費、保險、通關程序由誰辦理仍可能有多種不同的變化。例如國際商會（International Chamber of Commerce）的貿易條規（Incoterm）即為方便交易雙方約定而制訂了EXW到DDP等十一種規則。如果當事人並未在契約中明訂準據法與管轄地，債務履行地也可能影響準據法與管轄地的判

---

13 Restatement Second of Contracts § 302 "(1)Unless otherwise agreed between promisor and promisee, a beneficiary of a promise is an intended beneficiary if recognition of a right to performance in the beneficiary is appropriate to effectuate the intention of the parties and either (a)the performance of the promise will satisfy an obligation of the promise to pay money to the beneficiary; or (b)the circumstances indicate that the promise intends to give the beneficiary the benefit of the promised performance. (2)An incidental beneficiary is a beneficiary who is not an intended beneficiary."

斷。當然，為避免準據法與管轄地的判斷，受到各法域法律衝突法則與民事訴訟法之影響，建議當事人自己約定清楚為宜。

8. 履行的方法或標準

履行義務的方法標準當然也是看個別交易的類型，由當事人約定。今天，在全球的商務活動中有太多可以參照的品管標準、資訊安全標準、環安衛標準、社會責任標準。這些標準的細節內容不在本書的範圍內。英文契約中經常提及一些基本的履約標準，但締約各方可能不甚明瞭其真正的意函。例如「best effort」，或「reasonable effort」到底要求當事人要多努力？以下將簡要說明。

在英文契約中我們經常看到要求當事人要努力的條款，像是「best efforts」（最大努力）、「reasonable efforts」（合理努力）、「reasonable best efforts」（合理的最大努力）、「commercially reasonable efforts」（商業上合理的努力）、「good-faith efforts」（誠信的努力）、「diligent efforts」（勤奮謹慎的努力）、「commercially reasonable best efforts」（商業上合理的最大努力）、「good-faith reasonable efforts」（符合誠信的合理努力）、「every efforts」（盡一切努力）以及其他各種表示努力義務的條款。這些不同說法的努力義務在意義上有沒有差別？當事人到底應該要多努力？

在臺灣，有所謂的注意義務，可分為(1)善良管理人的注意義務，若違反屬於抽象輕過失；(2)與處理自己事務同一的注意義務，若違反則屬於具體輕過失；以及(3)一般人的合理注意義務，若違反則屬於重大過失。一般來說，善良管理人的注意義務程度高於，與處理自己事務同一的注意義務程度；而處理自己事務同一的注意義務程度高於一般人

的合理注意義務程度。但這也只是大概的標準，並沒有具體而明確的界線。

　　某家經營入口網站並提供網路行銷服務的業者，與線上學習經營者簽訂網路連結服務的協議。線上學習業者後來認為獲得的行銷服務效果不如預期，而終止契約。線上學習業者提出的理由之一，在於契約中提到「本合約期間，甲方應盡最大努力及資源安排，協助乙方課程商品廣告行銷及曝光，以利乙方課程商品之銷售及推動」。而且網路行銷業者在締約協商過程中，提供了多種有關電子報行銷、Email行銷，及其他行銷服務的廣告文宣。但法院認定在雙方口頭溝通的過程中，行銷服務業者提到除了提供入口連結以外，其他行銷服務是要個別商議評估。「尚難認被上訴人確有透過網站資源主動為上訴人提供其他廣告、促銷與宣傳等服務之義務。」[14]簡言之，法院不太看重「最大努力」這四個字。

　　但在另外一件訴訟中，臺灣廠商與外國軟體開發商簽訂軟體授權契約，契約約定授權廠商應「盡最大努力以去除任何經被授權人書面報告之重要偏差」，但授權廠商並未爭執其有無盡努力義務，而是爭執被授權廠商用電子郵件通知不算書面報告，且不符合契約中的通知條款。法院認定通知條款並未限制書面是什麼樣的書面，所以電子文件也算是書面。而後法院直接認定沒有除去軟體瑕疵就是違約[15]

　　在美國，有學者[16]從英文語意的觀點、以及案例法的觀點討論「最大努力」、「合理努力」等詞語的差異。就算是「最大努力」，用在一般情境中，也並不是「赴湯蹈火在所不辭」的意思，仍然是講得通的

14　臺灣高等法院99上易287號判決。
15　臺灣高等法院94重上更（一）97。
16　參閱註7引Adams著作第159-171頁。

努力。在商業實務上,一般共識是這些努力的差異只是感覺上的強弱。但感覺本身,當然是每個人都不一樣的。

美國的法院在幾個案件中認為「"Best efforts"...cannot mean everything possible under the sun...」(「『最大努力』……不是指太陽底下一切可能的方法」)(註17),「We have found no cases, and none have been cited, holding the "best efforts" means every conceivable effort」(「我們找不到案例,當事人也沒有援引任何案例認定『最大努力』就是各種想得到的努力方法」)(註18),「The requirement use its best efforts necessarily does not prevent the party form giving reasonable con-sideration to its own interest.」(「盡最大努力的要求,必然無法禁止當事人合理地考量自己的利益」)(註19)有的法院認為「最大努力」就是「act with good faith」(「按照誠信原則辦事」),有的法院認為「最大努力」要求的比「誠信」還多,有的法院用「reasonableness」(「合理性」)來定義「最大努力」,還有法院認為最大努力跟合理的努力是同一件事。

美國的統一商法典(Uniform Commercial Code)Section 2-306(2)本文提到「最大努力」,而評論中對「最大努力」的解釋是「to use reasonable diligence as well as good faith in their performance of the contract」(「在履行契約時,合理地勤奮謹慎並遵循誠信原則」)。此處「最大努力」跟「合理努力」似無區別。

英國法院有的認為「best endeavours」(「盡最大努力」)比「all reasonable endeavours」(「盡各種合理努力」)更努力,而「盡各種合

---

17 *Coady Corp. v. Toyota Motor Distributors, Inc.*, 361 F.3d 50, 59 (1st Cir. 2004).
18 *Triple-A Baseball Club Assocs. v. Northeastern Baseball, Inc.*, 832 F.2d 214, 228(1st Cir. 1987).
19 *Bloor v. Falstaff Brewing Corp.*, 601 F.2d 609, 614 (2d Cir. 1979).

理努力」又比「reasonable endeavours」（「盡合理的努力」）更強烈一些。不過這些法院仍然說不出個明確的標準。

澳大利亞有的法院認為最大努力還是要受合理性的限制，有的法院還是很想劃出不同的努力之間的分隔線。

從美國、英國、與澳大利亞的案例，我們很難期待用「最大努力」，法院就會要求義務人要更努力。

即使並沒有明確的標準說明各種努力條款到底要多努力，但美國的法院大多數還是承認這些努力條款的效力。伊利諾州法院是例外，該州法院認為這些努力條款太模糊了，除非當事人再講得更清楚點，否則努力條款有寫等於沒寫。還有一些法院認為，契約裡如果要求當事人要努力達成合意，這是不通的。因為生意喬不攏，並不能講是誰不夠努力。

努力條款，不管是最大努力、合理努力、最大合理努力、還是什麼努力，在美國大致上被認為是有效的，只是不見得能夠劃出明確的分隔線，從最弱的一端逐一排到最強的一端。

法國、比利時、巴西、埃及、英國、德國、義大利、荷蘭、波蘭、西班牙、瑞士、及美國的專家學者組成一個國際契約專家工作小組（註20），從西元1975年開始進行的系統性分析，其研究內容也包括「best efforts」、「reasonable care」、「due diligence」相關條款在國際契約上的運用。該項研究分析了經銷契約、營建契約、零組件製造供應契約、研發契約、技術支援契約、商標專利授權契約、衛星發射契約、letter of comfort、信用狀、相對貿易契約、公司購併契約中提及努力要求的部分。而努力要求經常出現在前述契約中的保密義務條款、良善管

20　Marcel Fontaine & Fillip De Ly, *Drafting International Contracts: An Analysis of Contract Clauses* 187-230 (Transnational Publishers, Inc. 2006).

理條款、不可抗力條款、降低損害的義務等。

該項研究將各類相關辭語分類成四組：

第一組要求義務人盡最大努力，包括有「best efforts」、「best endeavors」、「every effort」、「everything in its power」、「to the best of its ability」、「to the best of its experience」、「the best it can」及其他。

第二組則參照合理性的標準，包括有「all reasonable efforts」、「all reasonable means」、「reasonable care」、「all reasonably possible efforts」等。

第三組則提到勤奮謹慎的努力，包括「with due diligence」、「in a diligent manner」、「with utmost diligence and care」等。

第四組則引用業界相關標準，例如「Selon les règles de l'art」（「根據工藝標準」）、「in accordance with recognized professional standards」（「依據公認的專業標準」）、「in accordance with good engineering practice」（「依照良好的工程慣例」）、「entsprechend dem Stand von Wissenshaft und Technik」（「依據科學與技術的專業」）。有的契約則混用了不同分組的詞語。

雖然專家工作小組認為上述四組概念應該有不同的內容，也試圖藉由各國法院的判決加以類型化，不過最終仍無法整理出明確的概念定義。其對契約談判人員的建議，包括有參照締約當事人過去的行為、參照締約當事人的最大能力、參照專業群的一般標準，以及參照一般理性、謹慎的人的作為。

到底怎樣才算夠努力？法院可能有幾個參考點，其一是契約協商的時候，當事人到底說了什麼話；其二，是業界慣例、其三，是做出承諾

的人，過去在其他也要求要努力的契約裡，到底多努力；其四，是義務人與權利人如果兩家公司變一家公司時，義務人會多努力。

因此，學者建議就只用「合理努力」，對大家都好。想要用「最大努力」套對方當事人，一方面不一定有效，二方面試圖用含糊不清的條款，唬弄外行人，可能也有違律師執業倫理。當然在契約中如果要使用「合理努力」，可以在定義條款中再加上更明確的參考點，例如一般人的標準、義務人自己過去的作法、產業界的做法等。在協商努力條款時，可以舉些例子說明義務人該做到哪些事情。譬如經銷商要在哪些地區設立幾個辦公室，雇用銷售人員、客訴客服人員、參加當地重要的展覽、派人參加原廠訓練等等。當事人一方也可能想要設一些負面表列的排除條款，例如一年的廣告費用限額。所以，如何寫契約裡的努力條款，小結如下：

(1) 用「use reasonable efforts」或「make reasonable efforts」就好。

(2) 「efforts」通常是用複數。

(3) 加上努力的參考值，例如義務人自己過去是怎麼做的、行業標準、一般人合理的標準。

(4) 正面表列義務人該做哪些事情。

(5) 可以加上負面表列的排除事項。

9. 履行義務的時間

整個契約有其生效的時間，契約之下各項給付義務也可能有其履行的時間，例如付款期限、或交貨期限。這些時間是確定有到來的一天，

所以這裡講的時間，不同於條件。條件所指定的事件會不會發生，在締約時尚不可知。

時間可以用年、季、月、日、上午、下午、時等加以指定。在英文契約裡時間是個很容易造成困惑的問題，所以在第六章詳細說明。

10. 履行義務的條件

給付義務可以有附條件，例如買人壽保險，死亡時領死亡給付，屆滿一定年齡領回生存金。但能夠附條件的不只是給付義務。裁量選擇、宣示、擔保事項、甚至整個契約的效力都能受條件影響。因此有關條件的細節內容，將在後續章節詳細說明。

第三節　給付義務句寫作的其他相關考量點

1、盡可能用主動式，避免被動式

如果用被動式，首先，字數會增加，因為契約中要約定處理的事情很多，仍以精簡為要。其次，如果只強調什麼事情該被完成，常常一不小心就忘了交代誰要做，最後就是沒有人做。相對的，主、動、事，也就是主詞加動詞加受詞，這是大多數人最為習慣的構句方式。如果是禁止一方當事人做某件事，英文就用「shall not」或「may not」來表示。如果是各方當事人都不能做某件事，就用「neither party may」。

2、避免約定沒有義務做什麼事

　　本來，沒有義務就不用在契約裡廢話了，因為沒有義務的事情太多了，寫都寫不完。但某些情況下，可能法律推定有義務、或者因為當事人間的溝通過程，怕產生誤解，所以特別說清楚當事人一方沒有義務做某件事。沒有義務可以用「is not required to」或「is not obligated to」來表示，雖然這是被動式，但這裡還可以忍受。如果真得要避免被動式，也可以用「has no obligation/duty to」。但絕對不是在「shall」後面加「not」。

裁量選擇句與宣示句

第一節 裁量選擇句（Discretionary authority）

> ### 例句一：
>
> **The Bar's Right to Terminate Agreement for Material Breach of the Band.** If one or more Material Breaches of the Band Partnership occurs, the Bar Company may terminate this Agreement immediately by notifying the Band Partnership 3 days in advance.

　　例句一就是一個裁量選擇句，給付義務的反面就是受領給付的權利。但是裁量選擇跟受領給付的權利不太一樣。受領給付的權利通常是契約的主軸，像是受領買賣標的；但裁量選擇是較周邊的事項，像是解除或終止契約、發送通知的方法，將契約權利或義務移轉給第三人，認定違約、聲請拍賣擔保品等。(註1) 受領給付，通常必須看義務人臉色，所以義務人該做的事不做，或做了不該做的事，要用違約責任相繩。但裁量選擇看重的是權利人可以做什麼事情，權利人自己做了就做了，相對人連擋都擋不了。像是如果發生房東可以終止契約的情況，房東下達逐客令，講了就生效，一言既出，駟馬難追；房客也沒什麼辦法干擾讓通知不生效，所以不太需要考量義務人搗蛋時，怎麼用違約責任處理他。但是當然房東一通知終止租約，房客就要在期限前搬空，這搬空又變成一項給付義務，要看房客臉色，所以不搬空就要用違約責任處理。

1　Tina L. Stark, *Drafting Contracts: How and Why Lawyers Do What They Do* 29-31 (Wolters Kluwer, 2007).

1. 裁量選擇句的寫法

在英文契約裡，裁量選擇句的寫法很簡單，就是主詞加上may，再加上動作。主詞必定是契約當事人，其他人能做什麼，不能做什麼，並不是契約要談的。緊緊抓住may一個字就好，簡單明瞭，不要用其他任何字或加其他字，不要講「have the right to」、「shall have the right to」、「is entitled to」、「is authorized to」、「be free to」、「may … but is not required to」，或「A grant B the right to」。[註2] 同樣的，在裁量句裡也是寧可主動，不要被動，盡量用主動式來寫。

裁量選擇句時常也會加上條件。例如，房東不是隨便就能終止契約，要發生特定情況，例如房客連續兩個月不繳房租，經過催討還不繳，才能終止契約。

2. 明示其一，排除其他 v. 約所不禁，即為可行

在寫裁量選擇句時要留意，明示當事人一方可以做某件事，到底是說當事人只能做這件事，還是當事人可以做這件事，也可以做別件事。例如契約裡的紛爭解決條款寫「當事人就本契約所生的糾紛，得向台北地方法院提起訴訟以解決。」那能不能向高雄地方法院提起訴訟呢？或者在轉讓條款裡寫「建商得將本契約之權利與義務移轉由關係企業承受。」那建商能不能將契約權利義務移轉給非關係企業承受呢？當然可以查找契約所適用的準據法，不過在跨國交易中還是在契約裡寫清楚最直接了當。

2　Kenneth A. Adams, *A Manual of Style for Contract Drafting* 59-60 (2rd ed., ABA, 2013).

　　如果要明示其一，例外排除其他，就寫「當事人就本契約所生的糾紛，除臺灣臺北地方法院以外，不得向其他法院提起訴訟以解決。」中文的「除……以外」本身有些曖昧，意思可能是「besides」，也可能是「except for」，通常需要看下一句怎麼寫。「除臺灣臺北地方法院以外，不得向其他法院提起訴訟以解決」的「除……以外」是except for，也就是專屬的合意管轄。「除臺灣臺北地方法院以外，亦得向其他法院提起訴訟以解決」的「除……以外」是「besides」，也就是非專屬的合意管轄。當然我們可以講得更簡單「當事人合意以臺灣臺北地方法院為本契約所生一切糾紛的第一審專屬管轄法院。」

　　英文契約怎麼表示呢？此處我們必須想清楚我們到底要什麼。我們並不逼當事人一定要打官司，當事人可以自己談判，也可以找第三人調解，還有很多alternative dispute resolution。所以我們不能說「本契約發生任何糾紛時，當事人應向臺灣臺北地方法院提起訴訟」（If any dispute arises from this Agreement, both parties shall litigate in Taiwan Taipei District Court）。

　　有三種可行的做法，一種是轉換成不作為的給付義務句，再加上但書「當事人就本契約所生糾紛，不得向任何法院提起訴訟，但臺灣臺北地方不在此限」。（「Except for Taiwan Taipei District Court, neither party may bring a legal action or proceeding against any other party for any dispute arising out of or relating to this Agreement or the transactions it contemplates in any court」）第二種是用條件句，加上給付義務句來表示，「如果當事人擬提起訴訟，應向臺灣臺北地方法院起訴」。（「Any party bringing a legal action or proceeding against any other party for any dispute arising out of or relating to this Agreement or the transactions it contemplates shall bring the legal action or

proceeding in Taiwan Taipei District Court」）^{（註3）}第三種是先講當事
人可以做什麼，再限制但是應該如何做，例如「當事人得提起訴訟，但
應向臺灣臺北地方法院起訴」。（「Any party may bring a legal action
or proceeding against any other party for any dispute arising out of or
relating to this Agreement or the transactions it contemplates, but that
party shall bring the legal action or proceeding in Taiwan Taipei
District Court」）

　　有些人可能嘗試用「may ... only」來限制裁量空間，在某些情況
下，這麼做是可行的，不過留意將only緊貼著被限制的對象，放在被限
制的對象前面，而不是may的後面，其實在中文裡較不會發生這種問
題，例如「出租人**只能**依據第十條指定的方法通知終止租約」，用英文
說應該是，「Landlord may notify tenant of termination *only* in accordance with the methods stipulated in Article X.」，而不是「Landlord
may *only* notify tenant of termination in accordance with the methods
stipulated in Article X.」有時候用「may ... only」來限制裁量選擇是完
全不通的，這時候就用前面提到的三種解套方法來處理。^{（註4）}

　　而如果只是例示契約當事人的某項權利，就講清楚也可以做其他屬
性相同的事，例如「亦得向其他法院提起訴訟。」

3. 裁量選擇 v. 可能性

　　在英文裡may除了用來表示「可以做」，也可以用來表示「可能發

3　Brad S. Karp, *Governing Law And Forum Selection*, in *Negotiating And Drafting Contract Boilerplate*
　　140 (Tina L. Stark ed., ALM Publishing 2003).
4　參閱註2引Adams著作第62頁。

生」。所以在英文契約裡，盡量替「可能發生」找另外一個字來替代，有學者建議就用might。[註5] 此外，有時候「可能發生」是連提都不必提的，就直接刪掉。例如「在消費者物價指數發生任何可能的變動時，房東可以依照物價調整租金」（「For any change of Consumer Price Index that may happen, Landlord may adjust the Rent」），就改成「消費者物價指數變動時，房東可以依照物價調整租金」（「When Consumer Price Index fluctuates, Landlord may adjust the Rent」）。[註6]

4. 棄權（Waiver）與不棄權（non-Waiver）

棄權，顧名思義就是放棄權利。有些權利是契約給的，例如違約時有救濟權利，但有些違約情況並不是那麼重大，當事人可能認為維持關係比較重要，所以願意放棄救濟權利。這種情況是契約簽訂以後的事，所以此處暫且不談。

有些權利是法律給的，不是契約約定的。例如在美國，如果符合一定條件，訴訟當事人得要求由陪審團審判他們的民事糾紛；但也有許多人不喜歡找陪審團來審民事糾紛，所以要求對方在契約裡棄權。以下說明放棄要求陪審團審判的寫法。

---

5 參閱註2引Adams著作第63-64頁。

6 要讓租金跟著消費者物價指數調整，當然不能用這麼粗糙的方式約定，有基期設定，調整頻率等很多枝節的注意事項，可以參照行政院主計處，「訂定物價指數連動調整條款之一般原則」，http://www.dgbas.gov.tw/ct.asp?xItem=867&ctNode=2848, 最後點閱日：April 24，2011。

例句二：

Waiver of Jury Trial. Each party, to the extent permitted by law, knowingly, voluntarily, and intentionally waives its right to a trial by jury in any action or other legal proceeding arising out of or relating to this Agreement and the transactions it contemplates. This waiver applies to any action or legal proceeding, whether sounding in contract , tort or other wise.

（註7）

　　還有一種情況是權利人可能做什麼事、或沒做什麼事，就被法律當作是棄權了。例如對於一個事件可以行使幾種不同的權利，行使其一，是否就排除其他權利了呢？例如保險公司在投保責任保險的保戶通知出險後，有兩個選擇，一個是調查事件發生經過並為保戶辯護，另一個是抗辯該責任事件不在承保範圍內。保險公司如果選擇替保戶辯護，是不是就形同放棄抗辯該事件不在承保範圍內了呢？但如果不替保戶辯護，萬一保戶輸了官司，事後又沒辦法有效主張該事件不在承保範圍內，那保險公司就倒大楣了。所以保險公司通常會在契約裡明確聲明，保險公司調查事件經過並且替保戶辯護，不代表放棄事後抗辯承保範圍的權利。這就是不棄權的條款。

　　又例如，按照美國的不動產法，出租人即使在契約裡明確約定，要事先取得出租人的書面同意，承租人才能轉租或分租。但是依照法律原則，假使出租人第一次同意轉租，之後就被法律當做是放棄這道事先同意的關卡了，也就是轉租人再轉租出去，承租人也沒有不同意的餘地

7　註1引Stark著作第155頁。

了，除非承租人在第一次同意轉租的時候，就明白表示日後再轉租還是要取得事先同意。所以比較保險的做法，就是在租賃契約裡明白加上不棄權的條款，另外也打好書面同意的表格，表格也加上不棄權，日後轉租必須再經過事先書面同意的條款。不棄權的條款就是用締約當事人做主詞，加上 does not waive，再加上特定權利的說明。

第二節　宣示句（Declaration）

1. 宣示句的重要性

　　宣示句通常用在定義條款、準據法條款，及其他當事人選擇適用的政策。宣示句不創設給付義務、裁量選擇權限、也沒有聲明擔保的效果，所以不會有違反的效果，因為違反不了。一言既出、駟馬難追。所以給付義務、裁量選擇權限、與聲明擔保，又被稱為「operative languages」，宣示句就不是。但是宣示句是非常非常重要的。因為宣示句正是契約其他重要的句子（給付義務、裁量選擇、聲明擔保）如何做到輕薄短小，易讀易懂的關鍵。宣示句就像孫悟空的金箍棒，能伸能縮，伸展開來放在定義條款，縮起來放在其他各類句子作為 defined term，就可以看到完整的內容。就像下面例句三的「Tickets」，如果契約中每一個提到「Tickets」的句子都要把「admission certificates allowing holders to enter the Venue ...」這一長串寫出來，可以想見會是一場可怕的災難。

2. 宣示句的寫法

例句三：

Definition. As used in the Agreement, the terms defined in the preamble and in the Background section have their assigned meanings and the following terms have the meanings assigned to them in this Article.

(a) **"Tickets"** means admission certificates allowing holders to enter the Venue and observe Music Performances.

　　例句三是契約定義條款的部分，而「Ticket」帶出來的就是一個宣示句。契約一定都會提到許多不直接設定義務、裁量選擇權，而且也不是用來誘導當事人一方簽約的，甚至沒有對錯的問題。老闆跟員工可以約定工作時數是每週五天，一天八小時。不會因為臺灣的法定工時是兩週八十四小時，老闆跟員工這樣的約定就變成錯的。

　　所以如果有設定義務、裁量選擇權的效果，或者有誘導當事人一方簽約的效果，就不能寫成宣示句，而要用給付義務句、裁量選擇句、或聲明擔保句來寫。

　　宣示句通常用「A is B」，「A means B」，或者主詞加上其他現在式第三人稱單數動詞這樣的形式來表現。宣示句裡面不能出現shall、must、may這類表示給付義務、裁量選擇、或條件等具有法律效果的動詞。用宣示句來定義某個名詞時，如果所要做的定義已經周延，就用means，也就是說means就像一個等號。例如：

「**Closing**」 means the consummation of the transactions that this Agreement contemplates」[註8]

但如果定義的內容並不周延，就有點像「白馬非馬」的意思，白馬只是馬的一種，所以我們只能說「本契約所稱的馬包括白馬、黑馬」，就不能用「means」，而要用「includes」。「means」跟「includes」是勢不兩立的，所以在做定義時，不能同時用這兩個詞。若要排除棕色馬，就用「excludes」。

此外，必須留意，是就是，不是就不是，在英文契約裡不要用「is deemed to」。「is deemed to」指「不是的，也當做是」[註9]，也許在法規中有必要這樣做，但在契約中絕對不要給自己找麻煩，開個門讓對方以後可以拿反證來吵。當事人如果要約定「週末指的是星期五、星期六、星期日」就勇敢地這麼做，不要推定、不要視同、不要擬制、不要「is deemed to」。

宣示句也是契約基本句型當中，唯一不必然用契約當事人做主詞的句型。宣示句的時態要盡量使用現在式，即使不確定事情會在何時發生。避免用「will」等未來式表現。其實不需要擔心某些事件發生於未來，而非締約當時，用現在式好像怪怪的。因為閱讀契約條款的時機往往不只是締約當時，更重要的時點是發生糾紛的時候，當事人、律師、法官才會用放大鏡去看。更不能用「shall」，因為「shall」表示義務。[註10]

---

8 引自前註1 Stark 著作第 82 頁。

9 Tina L. Stark, *Drafting Contracts: How and Why Lawyers Do What They Do* 282 (Wolters Kluwer, 2007).

10 參閱註9引 Stark 著作第 31，32，144 頁。

英文契約的各項條款
與啟始部分

　　我們在第二章到第四章解說契約造句的各種句型，接下來，我們要說明契約的各種條款。句型就像是蓋房子的材料，水泥、鋼材、木材、塑料、玻璃、磚塊。各項條款就像是房子的各個構造部分，就像房屋有地基、有外牆、有屋頂、有門、有窗、有各個樓層、還有每一樓層的內部隔間。

第一節　起承轉合

　　起承轉合是東亞敘事的通用架構，無論是詩、文、小說、戲劇，作者多使用這樣的架構。英文契約長期發展出的通用格式，恰好也可以用起承轉合這樣的結構來統整掌握各種條款，如同以下表一的說明。

表一　契約各種條款匯整

起	標題、前言、敘文、合意文句
承	定義、主要義務、附隨義務（交割前後義務〔報告、協力、其他作為與不作為〕、保密、競業、廉潔、法令遵循、工具相關約定、教育訓練、技術協助支援、品質保固、驗證）、智慧財產權約定、聲明保證、效期
轉	終止或解除、違約救濟、履約障礙（不可抗力與艱困條款）、彌償（對第三人責任的移轉）、紛爭解決、債權移轉與債務承擔
合	通知、效力可分、完整合意、正本副本、簽名

　　以上表列只是大致情況，個別契約容或有項目增減。

　　英文契約的啟始部分，按照在契約文件中出現的順序有四個構成項目，也就是標題（Title）、前言（Preamble）、敘文（Recital），以及合意文句（Words of Agreement）。以下我們逐一介紹。

第二節　標題（Title）

　　一齣戲一定有一個題目，例如「山伯英台」或「牡丹亭」，由此提示觀賞者表演的大致內容為何，契約也是如此。一份契約最開頭就是標題，「名不正則言不順」，標題下錯了，很容易造成誤解。我們先看一個實例。

案例一

　　James C. DiPrima於M. B. D. Midwest, Inc.公司擔任總經理職務，1993年8月時，他代表公司簽訂辦公室租賃契約，同時也簽署了一份標題為"Personal Guarantee"的文件，在限定額度內擔保承租人會按時繳租，並在簽名欄上簽署"By: James C. Di-Prima"，而且註記職銜、公司全名與地址，以表明是以公司總經理名義，代理公司簽章，不是以自己名義簽章。James C. Di-Prima在三個月後即離職。四年之後M. B. D.公司在租期屆滿之前搬離，並積欠三、四個月的租金，房東把已經離職四年的James C. DiPrima告上法院，要求負保證人責任。James C. DiPrima抗辯說他簽擔保函是以公司名義簽署的。但法院認定"Personal Guarantee"這個標題中的"personal"有個人的意含，跟簽名欄以公司名義簽章有所歧異，所以准許原告援用契約之外的證據[註1]來證明James C. DiPrima實際上是以個人名義做保，最後判決James C. DiPrima敗訴。[註2]

1　英美契約法，如果契約內容完整而沒有曖昧不明的意義，就不能拿簽約之前的任何口頭或書面溝通過程來爭執，完全以最後簽訂的書面契約為準，這稱為parole evidence rule。

2　780 L.L.C v. DiPrima, 611 N.W.2d 637, 644-645 (Neb. Ct. App. 2000).

　　從以上個案可以瞭解，標題是很重要的。尤其不要故意下跟實質內容不同的標題，例如明明是銷售，為了稅務把標題寫成租賃。事後發生糾紛時，法院仍然會看實質內容，不會只看標題名稱。而且一個地方做假，往往假得不夠完全，得不到想要的效果，反而造成更多的困擾。所以誠實是最好的策略。

　　另外，在臺灣有印花稅，承攬契約要繳千分之一，買賣動產契約只要繳新臺幣12元定額，委任契約不用貼印花，所以契約的定性也會影響印花稅。如果標的價額很高，印花稅也不是小數目。若想要節省印花稅，不能只是亂改標題，還需要適當的分割或重組交易內容。

　　標題必須下得簡潔，又能夠提示這項交易的重要資訊。所以不要在標題裡寫「某年某月某日甲乙簽訂的一千雙男用皮鞋買賣契約」，因為這太冗長了。但也不要只寫「契約」兩個字。因為這看不出來到底是什麼契約。

　　在中文裡，我們說「買賣契約」，但寫英文契約時，我們寫「Agreement of Sales」就可以了，不需要寫「Agreement of Purchase and Sales」，有買就必然有賣。[註3]

　　對企業來說，契約的標題是契約管理上相當重要的環節，如果一家公司每年要簽訂數十份甚至數百份產品銷售契約，標題全部都寫「銷售契約」，日後要找跟特定公司簽訂的契約，簡直是大海撈針，如何找起呢？因此，標題可以寫契約的類別加上標的物的種類，並且置中，有人偏好所有字都用大寫不加粗體，有人偏好大寫加粗體。當事人的名稱可以放在第二行，然後加上括號，並且置中。例如：

3　Kenneth A. Adams, A Manual of Style for Contract Drafting 11 (3rd ed., ABA, 2013).

例句一：

PATENT LICENSE AGREEMENT
（FOR　　IN U.S.A., PATENT NUMBER: ）

第三節　前言（Preamble）

　　常常有人混淆前言（Preamble）與敘言（Recital or Background）。前言是標題下面的第一段，重述契約名稱並介紹當事人，這是必有的部分。敘言則是前言以下，說明交易背景資訊的段落，有時交易相當簡單，不需要敘言，也可能全部省略。但敘言還是有功能，留待後續說明。

　　再以戲劇比擬，傳統戲劇一開場，人物第一次上場往往是要交代自己的姓名，從哪裡來的，例如「小生姓張，名珙，字君瑞，本籍西洛人也」[註4]，不然觀眾哪裡知道這上來演的是誰呢？契約的前言就是這種交代的功用，交代什麼呢？前言主要再一次交代簽的是什麼約，簽約日期、締約當事人的身份，包括完整的姓名或組織名稱、在本契約裡當事人的簡稱、住址或營業所在地、組織的登記註冊地等資訊。

4　西廂記第一本第一章。

1. 前言的外觀形式

前言有多種寫法^(註5)。第一種寫法並非完整的句子，只是逐一說明相關訊息。

> ### 例句二：
>
> **Patent License Agreement**, dated April 5, 2015, between John Doe, domiciled at 1602 McClurg Ct., Chicago, Illinois 60611, U.S.A.（the "Owner"）, and ABC LLC., a Taiwan limited liability company（the "Licensee"）.

第二種寫法則是將這些訊息綴成完整的句子。

> ### 例句三：
>
> This **Patent License Agreement**, dated April 5, 2015, is between John Doe, domiciled at 1602 McClurg Ct., Chicago, Illinois 60611, U.S.A.（the "Owner"）, and ABC Corp., a Taiwan corporation（the "Licensee"）.

第三種則是分行列出相關資訊。

5　參閱 Tina L. Stark, *Drafting Contracts: How and Why Lawyers Do What They Do* 51 (Wolters Kluwer, 2007)，註3引用 Adams 著作第12頁以下。Sue Payne, *Basic Contract Drafting Assignments: A Narrative Approach* 373 (Wolters Kluwer 2011). Edward W. Daigneault, *Drafting International Agreements in Legal English* 60 (2nd ed. LPE, 2009).

例句四：

Name of Agreement: Patent License Agreement

Dated: April 5, 2015

Parties:

Owner: John Doe, domiciled at 1602 McClurg Ct., Chicago, Illinois 60611, U.S.A.

Licensee: ABC Corp., a Taiwan corporation.

第四種則是混合句子與表列方法。

例句五：

This **Patent License Agreement**, dated April 5, 2015, is between:

(1)　John Doe（the "Owner"）, domiciled at 1602 McClurg Ct., Chicago, Illinois 60611, U.S.A.; and,

(2)　ABC Corp.（the "Licensee"）, a Taiwan corporation with Uniform ID 21000000[註6].

2. 簽約動作

有人用「is signed and executed on」，有人用「is executed and

6　參閱 Tina L. Stark, *Drafting Contracts: How and Why Lawyers Do What They Do* 51 (Wolters Kluwer, 2007)，註3引用 Adams 著作第12頁以下。Sue Payne, *Basic Contract Drafting Assignments: A Narrative Approach* 373 (Wolters Kluwer 2011). Edward W. Daigneault, *Drafting International Agreements in Legal English* 60 (2nd ed. LPE, 2009).

delivered on」，看似「sign」、「execute」、「deliver」各有不同的意思。確實有古老的判決執著於不同字眼表示不同的意思，例如「sign」是自簽自受，「execute」是被授權簽名。[註7] 不過也有學者認為時代演進，已經不需要這種區別了，而且「execute」也有執行的意思，會造成混淆。

至於「deliver」，則確實與「sign」的意義不同。「deliver」表示交付簽好的契約。在英美法裡有一些契約，例如蓋印契約，房地讓渡契約、票據、載貨證券、倉單等權利證明證券等，除了作成書面並簽名以外，還必須交付。所以才會用「is executed and delivered on」。[註8] 雖然大部分的契約並不要求要交付，不過建議仍然寫明已交付（delivered）為宜。因為契約的精髓是意思表示合致，一定要互表合意，不能只是各表，各表各的，沒有拘束力。互表合意的重點，就是一方要向對方表示他的意思，對方也要回頭向這一方表示他的意思，而這兩個意思必須是一致的。在契約上簽名是為了表示意思，但是簽了約沒有交給對方，塞在抽屜裡，就沒有完成意思表示的動作，所以建議還是寫明契約已交付，避免日後發生糾紛。只是已交付的說明，可以寫在緊接契約簽名處之前較為適當，因為緊貼著簽名的地方，比較能證明確實已交付，如同以下例句六。而在前言中，只要寫「signed」或「dated」就可以了。

例句六：

IN WITNESS WHEREOF, the Parties have executed this

---

7 *Wamesit Nail Bank v. Merriam*, 96 A. 740, 741 (Me. 1916)，轉引自 Stark 前揭著第 188 頁。

8 參閱 Stark 前揭著第 188-189 頁。註3引用 Adams 著作第 122 頁以下。楊楨，《英美法契約論》（凱侖出版社，1984年），第 141-146 頁。

Agreement in two counterparts and delivered to each other.
　　Party A
　　By
　　Title

　　Party B
　　By
　　Title

3. 日期

　　簽約日期跟生效日期可能不同，如果是寫簽約日期，就用「dated」直接加日期。如果是寫生效日期，就用「dated as of」，然後在契約最末簽名處另行註記簽約日期。在契約中最好是分別寫出簽約日期與生效日期，簽約日期寫在前言內，生效日期寫在契約效期條款中。簽名處就不要再加日期，因為簽字的老闆可能常常忘記填上日期，或者怕填錯日期而乾脆不填，而且契約一定有兩方以上簽名，簽名的日期可能不一樣，為了避免好幾個日期愈看愈迷糊，就只保留前言的簽約日期跟效期條款的日期。

　　日期不要完全寫成數字式的例如「04／05／2015」，因為不同國家的人對日期的寫法有所不同，可能分不清楚上面這個日期，到底是四月五日還是五月四日。簽約日期跟生效日期當然可以不一樣，也不限制孰先孰後；但是目的要正當，要先想好稅務機關或證管機關會怎麼解讀，是不是存心詐騙稅務機關或投資人？

4. 當事人

(1) 前言中有關當事人的資訊

案例二

Liberty Mercian Ltd v Cuddy Engineering Ltd [2013] EWHC 2688 [註9]

原告簽了一份高台營建契約，為了建造高台以便將來蓋超市，標題是NEC3 Form of Contract。議約對象，自稱為Cuddy Group，事實上那是Cuddy Demolition and Dismantling Limited（CDDL）的商業名稱。契約草案中也寫簽約人是Cuddy Group。原告委託的律師請助理在網站上蒐尋資料，結果原告方的律師因為網站資料而認為Cuddy Group指的是另外一家公司Cuddy Civil Engineering Ltd（CCEL），原告方的律師主動要求Cuddy Group將簽約人名稱改為Cuddy Civil Engineering Ltd.全名。被告也不反對，就改了，事實上CCEL公司並未從事營業。後來發生糾紛，原告要求確認CDDL才是締約當事人，並主張違約強制履行與損害賠償。雖然之前有相似的案例，法院認定契約當事人名稱錯誤不影響契約仍在真正的交易當事人之間成立，但在本案中，法院認定是原告主動要求變更締約當事人名稱的，所以締約當事人就是CCEL。

在第一章裡，我們提到契約的四個重點，第一個就是跟誰交易。

當事人的資訊必須詳盡而能確定到底是跟誰交易。作者在工作中常

9　https://www.casemine.com/judgement/uk/5a8ff73660d03e7f57ea9b66，最後點閱日：July 5，2023。

常看到契約裡面當事人資訊不詳細明確的情況，這可能產生的嚴重後果，例如發生糾紛時，找不到被告。當事人如果是法人，必須列出的資訊包括全名、組織屬性、登記地址、編號（在臺灣是統一編號）、登記的法域。當事人如果是自然人，必須列出的資訊包括全名、國籍、本國人戶籍地址（外國人則提供住所地）、身分證字號（外國人則列護照號碼，在臺灣是統一編號）。

我們要特別說明的，簽約的對象必須是人。這點聽起來很奇怪，難道有人跟不是人簽約嗎？所謂的人，在法律上有自然人與法人。自然人就是你我這些活生生的人，法人就是法律賦予獨立人格可以用自己的名義進行法律行為的人，有任何糾紛就找那個法人，而不找法人背後的自然人。

實際上常常看到跟不是人簽約的情況，發生訴訟的時候，才發現跟自己簽約的不是人。舉例來說，在臺灣如藥局、診所、商號（通常名稱是某某企業社、某某行）、合夥、工廠都不是人，所以如果跟上面這些事業體簽約，卻沒查清楚它的負責人是誰，打官司的時候可能找不到對象，就算它有統編，也不能作為有效的被告。法人常見的有公司、財團法人（例如財團法人某某基金會）、社團法人某某協會。法人自己有獨立的法律上人格，但藥局、診所、商號、合夥這些，則沒有獨立的法律上人格，他們不能簽約，只有他們的負責人才能簽約，所以如果跟某某藥局簽約，需要查明負責人是哪一個藥師，法律上就是跟那位負責人簽約，要寫在契約中，避免遺漏，找不到人。

另外，以法人的名稱來說，如果只寫簡寫，又沒有統一編號或公司註冊編號，同一個地址可能登記多家開頭部分名稱相同的公司，最後可能根本找不到訴訟的對象。

　　一般來說，在同一個法域（jurisdiction），(註10) 事業名稱不能完全相同，所以識別事業有效的方法就是事業名稱加上組織屬性與法域。

　　查清楚交易對象的法人全名之後，建議還需要查詢該法域的公司登記資料，才知道公司資料有沒有錯誤，或者只是假名、或者已清算解散、未繳年費而被除籍，甚至是虛構的公司。在美國，有所謂的d/b/a（doing business as），必須特別留意。d/b/a就是做生意的假名，有些公司簽約故意用d/b/a，不用真實的名字，有的d/b/a是有登記的，還有一些可能並未完成登記，如果發生訴訟，可能影響被告身分的特定，建議務必查清楚，用真名簽約。作者工作上常發現有些美國公司，同一批人故意開設多家開頭名稱相同的公司，只是各公司組織屬性不同，有的是有限公司，有的是股份有限公司，甚至交叉使用其他公司的正式名稱作為d/b/a。各個國家的組織屬性分類都不相同，有的像臺灣這樣分成有限公司、或股份有限公司，有的則是區分公開公司或私人公司，公開公司指的是股票在證券交易所中買賣的公司，私人公司則是股票不上市的公司。

　　像案例二這種搞錯簽約對象的情況非常悲慘，所以簽約時，不要自己猜測對方是誰，如果對方的名稱不完整、不明瞭，直接要求對方提供公司完整全名，再依據對方的資訊進行查核與徵信。

(2) 當事人的簡稱

　　針對當事人的簡稱，我們特別說明，契約中常見幾類當事人的簡稱，第一種是「甲方」、「乙方」、「Party A」、「Party B」、「First Part」、「Second Part」。我們建議不要用這種簡稱。雖然這種簡稱容易

10　一般來說，法域是一個國家，但在美國，每個州都是一個法域。

複製貼上，但是連寫契約的人都可能搞混甲方乙方，而且這樣的錯誤很難一眼看出來。如果甲方是買方，乙方是賣方，契約條款寫甲方應交貨，乙方應付款，很難一眼看出問題。

第二種是用個體屬性或名稱簡寫，例如「公司」（「Company」）、「合夥」（「Partnership」）；或名稱縮簡寫，例如「ABC」，「DEF」。公司名稱縮簡寫沒辦法複製貼上，每一份契約都要重新修改「ABC」，「DEF」這種縮簡寫。而公司或合夥，如果交易雙方都是公司的時候，「公司」到底是稱呼哪一方容易產生混淆。而且跟甲方乙方一樣，寫錯了不容易發現錯誤。

第三種則是用交易中的角色來稱呼，例如「Seller」、「Buyer」。這種簡稱既容易複製貼上援用，寫錯時又容易發現錯誤，所以作者推薦採用。但要注意英文中有很多交易角色的名稱常常只差一兩個字母，像是「licensor」（授權人）、「licensee」（被授權人）、「mortgagor」（抵押人）、「mortgagee」（抵押權人）、「lessor」（房東）、「lessee」（房客），要盡量避免使用這些字，因為即使是以英文作母語的人都不一定搞得楚，更何況是我們。所以盡量用別的字替代，例如「borrower」、「lender」、「landlord」、「tenant」。

如果對交易對象很不熟悉，可以考慮是否可以要求對方提出保證人，以提升信賴，或者透過第三方機構的相關機制，例如價金或標的信託等履約保證機制，以降低風險。

(3) 三個以上該用「between」還是「among」？

締約當事人如果只有兩方，用「between」當然是沒問題。但締約當事人有三方以上的時候，就有人擔心寫「between」不對，而改成

「among」。其實不論締約當事人有多少，「between」才是正確的。兩方或兩方以上之間建立「直接的」相對關係都是用「between」。「among」表示群體內彼此之間較為「間接的」關係。(註11)

第四節　敘文（Recitals）

　　敘文就是標題、前言（當事人介紹）之後的下一段，這段主要交代交易的背景緣由與事實。也有人稱直接稱這段為背景（Background）或鑒於條款（WHEREAS Clause or Witnesseth），因為常見的格式就是「WHEREAS xxxxx, WHEREAS xxxxx, WHEREAS xxxxx」。

案例三

Wood v. Lucy, Lady Duff-Gordon (註12)

　　被告Lucy Duff-Gordon是個服飾設計師，聘請原告Otis F. Wood行銷她所設計的服飾，並授予原告至少一年的獨家銷售權。原告與被告並約定分享銷售獲利。契約還約定自動更新，除非一方在九十日前通知他方終止。

　　在大陸法系，這樣的契約大概沒有問題。但在英美法系，這樣的契約問題可大了。因為時尚服飾設計師給銷售商一年獨家銷售權，銷售商拿什麼來換？沒有明確的對價關係，也就是沒有約因，從而契約就變成沒有拘束力。被告就是這麼抗辯。但法官Cardozo從雙方契約敘言的內容、分享利潤的機制、以及原告服飾銷售商承諾幫被告設計師視情況需要申請專利權、著作權與商

11　Stark 前揭著第54頁，註3引用 Adams 著作第18頁。
12　118. N.E. 214 (N.Y. 1917)

標權，認定原告確實提供了對價，所以雙方契約有拘束力。契約敘言是這麼寫的 "the said Otis F. Wood possesses a business organization adapted to the placing of such indorsements as the said Lucy, Lady Duff-Gordon has approved"。法官認為這表示原告的事業組織為了達成契約的目的而做了一些調整。

1. 敘文的功能

敘文雖然沒有執行力，但絕對不是雞肋。敘文有許多的妙用。

例如，在商人對商人的貨物銷售契約，依據美國統一商法典（Uniform Commercial Code）第2-312條第3項規定如下：

「除非雙方另有合意，賣方如果是常業經營該類產品的商人，即擔保所交付的產品不會有第三人合法主張侵權或類似主張，但買方如果提供規格給賣方，買方必須填補賣方因為遵循該項規格而受第三人求償所生之損害。」

所以我們可以看到有些契約的敘文就會把前述條文產生擔保的前提條件寫入敘文，以引導解釋成具有前述之擔保，例如「賣方係常業經營該類商品的商人。」（The Seller is a merchant regularly dealing in Products）或者反面，賣方為了不做此項擔保，並要求買方填補其未來可能因第三人求償而生之損失，而在敘文中記載「買方提供規格，希望賣方供應依該規格製造之產品。」（The Buyer furnishes specifications to the Seller）。總之，把前提要件之成立記錄下來，若整個契約確實能適用某一州所採用的統一商法典，就會產生特定的效果，而無須將效果約定在契約中，懂的人就知道鋪排這種前提要件事實的後果，不懂的人

則不知道記錄這樣的事實會產生何種後果。

又例如，在商人對商人的貨物買賣契約裡，如果賣方不願意明確擔保產品的某項功能，但在敘文裡寫明，賣方是以銷售契約標的為常業的商人，即可能支持默示擔保（implied warranty）中適銷性擔保之存在。例如美國統一商法典第2-314條規定如下：

「2-314：適銷性；商業習慣

(1) 除非明文排除或變更（見第2-316條）如果賣方是常業從事該類產品銷售的商人，銷售契約即默示具有產品具適銷性之擔保。於本條中，有償提供食物或飲料，無論內用或外帶，均屬銷售。

(2) 具適銷性之產品必須至少符合以下要件：

    (a) 在行業內一般對於契約中有關該產品的描述並無異議；而且

    (b) 在種類之物，具有符合該描述內容產品之允當平均品質；而且

    (c) 適於該等產品通常用途；而且

    (d) 在合意容許的變動範圍內，各個相關部分在種類上、品質上與數量上屬於均等；而且

    (e) 依合意內容提供適當的容器、包裝，與標籤，而且

    (f) 容器或標籤上如有任何允諾或事實確認，均與實際情況相符；而且

(3) 除非明文排除或變更（見第2-316條），可能因為交易往例或商業習慣而產生其他默示擔保。」

所以在敘文中常看到記載「賣方是常業從事契約產品銷售的商人」

（the Seller is a merchant with respect to Products）。如果合約約定適用美國某一州作為準據法[註13]，則很可能適用統一商法典，如果擔心在契約上白紙黑字寫出適銷性擔保會引發對方抗拒反制，甚至對方反而要求明文排除，則買方在敘文中加入這樣一句話，即可能產生期望的適銷性擔保效果。

此外，買方商家之所以購買某項產品，很可能是為了某個特定用途，而且還跟賣方詳細討論過，經由賣方推薦此項產品後，雙方簽訂買賣契約。事後如果產品效用个如預期而產生糾紛，或許就能依據法律主張適於特定用途的擔保。例如美國統一商法典第2-315條規定如下：

「2-315：默示擔保：適於特定目的，如果賣方在締約當時有理由知道產品需求的特定目的，而買方倚賴賣方的技術或判斷以挑選或組裝適合的產品，則除非依據次條規定明文排除或變更，則有產品適於該特定目的之默示擔保。」

也因為上面這樣的規定，所以在敘文中我們可以看到老練的律師，會記載「買方仰賴賣方的技能與判斷以挑選或組裝適當的產品」。

又假設兩家公司談妥交易，中間撮合的人可能是其中一家公司的董事，而且這名董事可能跟交易雙方都有利害關係，按照紐約州的法律，[註14]如果這名董事沒有對公司說明自己在對方公司的利害關係，這項交易事後可能被撤銷。但是如果交易當事人能證明，這項交易各方面的條件對雙方都是公道而合理的，就不能撤銷。有時候中間撮合的人願不願意對自己的公司開誠佈公，並不是交易對手能掌控的。所以在這種情況下，如果契約的敘文寫明，這項交易協商的過程中，市場情況如何，找了哪些專家來做鑑定估價，定出來的條件對雙方都很合理公道，就能

13　路易斯安那州與波多黎各並未全面採用統一商法典。
14　New York Business Corporation Law § 713.

避免日後不必要的困擾。

又例如，買方跟美國的大盤商買一批貨（農產品除外），結果那批貨可能設定了動產擔保，如果賣方本來就是做銷售這一類商品的生意，而買方就算知道這批貨有設定擔保，但不知道在擔保的條件之下，賣方不能賣，買方在一般的業務往來中（Buyer in the Ordinary Course of Business）買下這批貨，按照美國統一商法典（Uniform Commercial Code），買方還是受法律保障的，有擔保債權的債權人只能自己跟債務人解決。[註15] 所以一般業務往來的場景就可以在敘文中加以設定。

敘文如何巧妙運用，當然必須視個別交易內容而定。但我們可以將敘文簡要歸納出幾類主要功能。

第一種功能，是支持、強化本來可能被質疑的契約效力，例如限制競爭的契約可能不容易過關，但是如果是把一家公司所有的營業資產賣給另外一家公司，那賣方股東簽訂限制競爭條款是很正當的，也較容易被法院接受。

第二種功能，是建立解釋規則，[註16] 讓法院之後介入時，知道雙方這項交易的大概情況，如果契約條款寫得有些模糊，至少法官會有個方向。

第三種功能是，釐清對待給付關係，[註17] 在英美法下，欠缺約因（consideration，也就是對價關係），可能整個契約就無效了，要不要履行，完全看義務方守不守信用。美國許多州採用英國留下來的防止詐欺法（Statute of Frauds），按照防止詐欺法的要求，某些交易必須有書面，才能由法院要求當事人履約或向法院主張賠償，這些交易包括有關

15　U.C.C. §9-320（a）

16　參閱 Marcel Fontaine & Filip De Ly 前揭著第88-89頁

17　參閱 Stark 前揭著第61-62頁。

土地利益的交易、無法在一年內完成履行的交易、五百元以上的商品交易，擔任保證人、以結婚的允諾換取對價、遺產執行人或管理人個人承擔遺產債務的允諾等。而且，在這些交易的書面契約裡必須列出約因。因此，可以在敘文中簡要說明雙方重要的對待給付關係。

第四種功能是，確立重要事實，以塑造出對我方有利的規範模式，例如前面提到有利害關係的董事居中牽線促成交易，契約就要寫清楚相關交易條件是很公道的，就算對方公司發現他們的董事沒交代利害關係，也不影響我方權益。或者為了支持默示擔保，而記載第五種功能是，應付政府，有時候政府機關、或甚至跨國組織會設定一些規範要求，要符合這些要求才有優惠或補助，或至少讓政府機關比較願意高抬貴手開放一些管制規定。所以契約裡面偶而還是需要呼個口號。例如上世紀七〇年代一些開發中國家在聯合國裡推出的「新國際經濟秩序」提案（New International Economic Order）[註18]。

第六種功能是，作為當事人間折衷妥協的辦法，也就是各方對特定的權利義務、聲明擔保或契約生效、失效條件無法達成共識，就在敘文中做可接受的表述。例如把締約的背景事實寫出來，如果背景事實認知有誤，雙方事後或許可以用重大的認知錯誤為理由撤銷契約，或者情事變化，造成履約目的受挫，也有解除、終止或變更的空間[註19]。

第七種功能，是建立禁反言（estoppels）的情境[註20]，在大陸法系，當事人往往可以拿契約文字以外的溝通內容，來爭執契約文字不代表當事人真正的真意。但在英美法的口頭證據排除法則（parole evidence）之下，只要契約文字寫得很清楚明確，而且當事人已經明白約

18　Wikipedia, *New International Economic Order*, http://en.wikipedia.org/wiki/New_International_Economic_Order, latest accessed on 2011/5/21.　Marcel Fontaine & Filip De Ly 前揭著第 67 頁。

19　參閱 Marcel Fontaine & Filip De Ly 前揭著第 74-79,91-94 頁。

20　參閱 Marcel Fontaine & Filip De Ly 前揭著第 93 頁。

定所有的交易條件內容都整合寫進契約，則在簽約同時或簽約之前任何的口頭或書面溝通內容都形同作廢了。所以在跨國交易中，把重要的事實寫進敘文，可以緩和口頭證據排除法則的不利影響。

第八種功能則是，把敘文當故事摘要，讓公司經理人及業務承辦人員可以藉此瞭解交易內容。

2. 敘文的內容

敘文有數類內容，包括當事人的特質屬性、交易的背景環境、談判的過程、締約目的、動機與基本精神、當事人除本份契約以外同時或先後簽訂的相關契約文件、當事人一方重要的陳述，以及交易雙方主要的對待給付內容[註21]。

敘文本身是沒有執行力的，它只是契約的附屬物，而非契約本體，所以如果影響權利義務或責任的事項，就不要只放在敘文裡面，而必須寫進本文。如果雙方在本文協商過程中對特定的權利義務或聲明擔保僵持不下時，敘文裡的折衷表述也是一種解決方法。當然，把契約本體裡的重要事項在敘文中重述一次，並不會影響這些重要事項的效力，不過要考慮有無必要，以及會不會多寫多錯，寫的前後不一。敘文與契約本體有不一致的情況時，有兩個判斷原則，(1)寫清楚的比曖昧的效力優先，(2)契約本體優先於敘文[註22]。

21　註3引用 Adams 著作第29頁。Marcel Fontaine & Fillip De Ly, *Drafting International Contracts: An Analysis of Contract Clauses* 63-79 (Transnational Publishers, Inc. 2006).

22　*Jamison v. Franklin Life Ins. Co.*, 136 P.2d 265, 269 (Ariz. 1943), quoting *Williams v. Barkley*, 58 N.E. 765, 767 (N.Y. 1900), quoting English case law (citation omitted). 轉引自 Stark 前揭著第63頁。

3. 敘文的寫法

敘文其實不需要標題，如果真的要替敘文下個標題，其實不適宜寫

「WITNESSETH」，這是古文，契約是給現代人看的，可以直接寫「Recitals」或「Background」。

敘文本體的起始，也不要用「WHEREAS,」，因為這同樣是兩百多年前的寫法了。敘文就用簡單的文字，以敘事體說明真正重要的締約目的，重要的背景事實，或當事人一方的陳述內容。

前言或敘文裡的專有字詞，必須在前言或敘文中就加以定義，不要讓讀者往下找定義條款才能知道專用名詞的意思[註23]，契約中如果有定義條款，通常也會在該條款說明定義的對象，不包括前言與敘文的字詞。但是前言或敘文的定義都不是用宣示句[註24]獨立造句的，而是用嵌入式的，因為宣示句較長，放在敘文裡會讓整個行文不是那麼順暢。也就是在其他句子中用括號內容補充說明，對某項事物給予某個名稱。例如例句七「Each party hereto（the "Receiving Party"）has reques-ted certain confidential information and proprietary information of the other party（the "Disclosing Party"）for the purposes of evaluating a transaction between the parties（the "Transaction"）」出現了三個嵌入式的字詞定義。但有時候真的沒辦法使用嵌入式定義立即說明時，也可以直接用第一個字母大寫帶出被定義的字詞，然後加上括號，說明這個專有字詞的定義在定義條款第幾項可以找到。例如，範例六最後一句的「this Agreement（as defined in Section 1.1）」。

23　Stark 前揭著第 64 頁。

24　例如，「『工作日』指每個星期的星期一到星期五。」宣示句的句型結構就是「被定義的詞」，加上「指」，加上「定義內容」。

4. 敘文範例

例句七：

Background. [註25]

Each party hereto（the "Receiving Party"）has requested certain confidential information and proprietary information of the other party（the "Disclosing Party"）for the purposes of evaluating a transaction between the parties（the "Transaction"）. As a condition precedent to providing any such confidential or proprietary information and continuing to negotiate the terms of the Transaction, Disclosing Party requires Receiving Party to execute this Agreement（as defined in Section 1.1）and abide by the terms hereof.

第五節　合意文句（Words of Agreement）

合意文句就是敘言之後簡單地說明雙方合意本契約的內容，舊式的寫法如下：[註26]

---

25 International Trade Centre UNCTAD/WTO, *Confidentiality Agreement (Example)* 200/2/1, http:// www.jurisint.org/doc/orig/con/en/2001/2001jiconen73/2001jiconen73.pdf, latest accessed on 2011/5/22。原示範契約將本項列為契約第一條，此處將它改為敘文使用。

26 Stark 前揭著第 65-66 頁。

例句八：

NOW, THEREFORE, in consideration of the premises and of the mutual agreements and covenants hereinafter set forth, the Owner and the Licensee hereby agree as follows:

除非老古板的老闆或當事人一定要你這麼寫，否則不用這麼囉唆地寫這種舊式的合意文句。舊式的合意之所以要寫「in consideratiōno」，是因為英美契約法要求，要再交代一次約因。當然在敘文的部分建議還是依循傳統寫一下約因。但合意文字就可以不用再講約因了。

新式的寫法如下：

例句九：

Accordingly, the parties agree as follows:

英文契約的定義條款
（Definitions）

定義條款屬於起承轉合中的承。

契約中的字詞[1]定義可能分散放在其他條款中（例如在不可抗力條款中，說明何為不可抗力），也可能集中成為定義條款（例如將產品、營業日、智慧財產權、機密資訊等各種詞語定義通通集中在契約的某一條，如果是英文契約，即按照英文字母的順序排列字詞定義）。

定義條款通常是契約的第一條，接在標題、前言、敘文與合意文字之後。但是，這不表示在撰寫契約時，優先撰擬定義條款。有時候反而先把權利、義務事項做完之後，再加上定義條款會比較方便[2]。

第一節　集中式的定義與分散式的定義

如果是集中式的定義，也就是在第一條提供定義，寫法就是運用宣示句來造句。如果是分散式的定義，則有兩種寫作方法，一種仍然是用宣示句獨立造句。另一種則是嵌入別的句子當中，例如：

例句一：

During the Term, the Employee shall work at places designated by the Employer on Monday through Thursday（the 'Business Day'）weekly。

在例句一中，「Business Day」的定義就被嵌入給付義務句。嵌入的方法就是在被定義的字詞後面加上括號，括號中置入定冠詞與引號加簡稱。但嵌入的定義方法有時會產生問題，例如以下案例：

1　雖然被定義的大部分是名詞或名詞詞組，但偶爾也需要對動詞或其他字詞做定義。
2　Edward W. Daigneault, *Drafting International Agreements in Legal English* 63 (2nd ed. LPE, 2009).

案例一

G. Golden Assocs. of Oceanside, Inc. v. Arnold Foods Co., Inc.（註3）

1983年8月某食品研發公司Golden Associates提供製造皮薄酥脆的法式麵包的技術資料給某食品製造商Devonsheer，由Devonsheer負責製造與行銷，然後與Golden Associates分享利潤。過了兩年，這項麵包的銷售情況不佳，Devonsheer將這項技術授權與行銷的契約轉讓給另一家食品公司Arnold。Arnold又做了幾年這項麵包生意，之後在1987年完全停止這項麵包產品的生產。在1989年Arnold的母公司CPC買下另一家也製造法式麵包的公司。Golden Associates起訴告Arnold違約、不當終止契約，以及不當干擾契約關係。這件糾紛的爭點之一在於契約敘言提到：

WHEREAS, Seller possesses technical information and know-how（the "Technical Information"）relating to the production and manufacture of food products which look similar to the thin, crispy crust of 'french bread' from which the dough has been removed（the "Products"）and is the sole owner of the trademarks "La Crunch Une" and "La Crunch" which it has used in connection with sales of the Products.

原告主張上面這句話所稱的 "Products" 是指 "food products which look similar to the thin, crispy crust of 'french bread' from which the dough has been removed"。所以只要看起來像是薄脆的法式麵包，就是雙方技術授權的標的，被告做任何銷售都要付

3　870 F. Supp. 472 (E.D.N.Y. 1994).

> 錢給原告，被告則主張敘文裡面又提到技術資訊、又提到產品，所以要連起來看，只有用到原告技術資訊的麵包，才要算錢給原告。

　　從以上這個案例，我們可以看到嵌入式定義的毛病，就在於難以分辨括號中的定義到底射程多遠。所以，若要使用嵌入式定義，需特別留意如何避免此種困擾。

　　以下則是集中而且非嵌入式，也就是獨立式的定義例子：

例句二：

Article 1 Definitions

1.1 Definitions. Terms defined in the preamble and the recitals of this Agreement have their assigned meanings, and the following terms have the meanings assigned to them [註4]：

(a) **"Breach"** includes a cross-default" [註5]

(b) **"Business Day"** means Monday through Friday, except for days on which commercial banks in Taiwan are authorized by laws or government instructions to close.

(c) **"Notice"** has the meaning assigned in Section 12.04.

(d) **"Patent License Agreement"** means the Patent License Agreement, dated May 20, 2020, between ABC Co. and DEF. Inc., as amended from time to time.

(e) **"Telephone"** means an instrument for reproducing sound

---

4　Tina L. Stark, *Drafting Contracts: How and Why Lawyers Do What They Do* 79(Wolters Kluwer, 2007).

5　Stark 前揭著第80頁。

at a distance [註6] , but excludes a cellular telephone that takes photographs.

1.2 Interpretive Provisions. [註7]

(a) The words "including," "includes," and "include" are deemed to be followed by the words "without limitation".

(b) References to a Person include that Person's permitted successors and assigns and, in the case of any governmental Person, the Person succeeding to the relevant functions of that governmental Person.

(c) All references to statutes and related regulations include

(i) any past and future amendments of those statutes and related regulations; and

(ii) any successor statutes and related regulations.

(d) All references in this Agreement to "Dollars" or "$" refer to lawful currency of [Country Name].

(e) All references to a time of day are references to the time in [location].

(f) If any date specified in this Agreement as the only day, or the last day, for taking action falls on a day that is not a Business Day, then that action may be taken on the next Business Day.

(g) The word "shall" is used to express a mandatory duty.

(h) The word "may" is used to express discretion.

(i) The word "must" is used to express condition.

(j) The word "terminate" means to put an end to this

6　Merriam-Webster's Collegiate Dictionary 1211 (10th ed., 1999)，轉引自 Stark 前揭著第 81 頁。

7　Stark 前揭著第 87-88 頁。

> Agreement by which all obligations still executory on both sides are discharged but any right based on prior breach or performance survives [註8].
>
> (k) The word "rescind" means to make void this Agreement.

第二節　定義條款的細節說明

1. 引言

　　定義條款要加上引言，說明定義條款所要定義的對象是出現在契約本文的字詞，原則上不包括前言與敘文中的字詞。所以前言與敘文中的字詞要先用嵌入式定義加以說明，真的無法嵌入時，就在專有字詞後面加括號說明，在定義條款中的第幾項找得到定義。

2. 完全定義

　　契約字詞的定義方法有完全定義、部分定義、跟反面定義。完全定義指的就是被定義的字詞完全等於後面定義的內容。這裡的參照標準完全就是契約的內容，而不管商業習慣或其他標準。例如，我們可以定義「『營業日』是星期一、星期二、星期三與星期四」。只要在契約中，當事人的意思如此，營業日就是這四天，不用擔心在外面的世界，公司行號在星期五也營業。

8　U.C.C. § 2-106(3)

在英文契約裡，提示完全定義的動詞就是「means」。例如：

例句三：

"**Business Day**" means Monday through Thursday.

在定義條款中，被定義的字詞開頭字母要大寫並粗體。但是在其他各條出現定義字詞，就只要開頭字母大寫，不需要粗體。

3. 部分定義

部分定義，用於提示某件事特別言明，所以納入定義範圍之內，若不特別言明則是否屬於定義範圍內，即不明確。例如貸款契約中常有的交叉違約條款。借款人A向B銀行借錢，B銀行可能在契約中要求加入交叉違約條款，效力就是A如果欠C錢到期沒還，也算是A對B違約。本來A欠C錢，到期不還，是A與C之間的事，但B怕自己變成最後一隻老鼠，別的債權人先把剩餘財產分完了，所以要特別約定A欠別人錢到期不還，也是A對C違約。

英文契約中提示部分定義的動詞就是「includes」。例如：

例句四：

"**Breach**" includes a cross-default.

4. 反面定義

反面定義，就是把特定事項排除。完全定義可以跟反面定義併用。例如：

例句五：

"**Telephone**" means an instrument for reproducing sound at a distance, but excludes a cellular telephone that takes photographs.

要特別留意，若使用完全定義，並舉例，又加上限定詞組，再加上反面排除，真的很容易造成災難。第一，想清楚這麼做的必要性；如果沒有必要，就精簡一點。第二，如果真的需要這麼做，順序上先限定、再排除。也可以用列舉（enumeration）的方式（也就是在「such as」、「that」、「excludes」之前加上「(1)」、「(2)」、「(3)」），或是用表列的方式（tabulation）的方式讓句子更加明確，避免誤解。

5. 參照定義

如同之前所述，字詞的定義可以集中放在定義條款，也可以分散擺在其他條款。如果在不可抗力條款中才定義什麼是不可抗力，在定義條款中同樣要列出簡稱，並參照實際定義的條號。例如：

例句六：

Section 1 Definition.

...

"Notice" has the meaning assigned in Section 12.04.

當然，在契約第十二條第四項就要對通知做定義，例如：

例句七：

Section 12.04

(a) Definition. "Notice" means a written statement of facts
for the events designated in this Agreement.

(b) ...

所以如果很多條款都會用到同一個字詞，就把它放在定義條款中，
而不要在其他個別條款才做定義，免得要交代參照很多條，反而麻煩。
另外，前言與敘文中的定義不需要在定義條款中做參照，因為我們在定
義條款的引言中已經明白表示，定義條款所定義的字詞不包括前言與敘
文中的字詞。

6. 詮釋條款

有些契約會在字詞定義之後加上詮釋條款。詮釋條款的功能在於說
明契約中出現的動詞、通貨、計量、計日或計時方法。這些詮釋條款通
常是被定型化，可以廣泛使用在各種不同的契約。例如「shall」的用

法，雖然美國的主流意見認同「shall」是表示義務，但也有人主張應該使用「must」。在英國、加拿大、澳洲、紐西蘭及其他英語國家，可能又有不同的習慣。因此，在詮釋條款中明確界定是較為穩健的做法。

7. 定義條款其他注意事項

(1) 不要在定義裡加入權利、義務、或聲明擔保事項

　　這是最為嚴重的，也是很常出現的錯誤。發生糾紛的時候，對方很可能爭執定義條款根本沒有設定權利義務或聲明擔保的效果，或者重要事項故意藏在定義條款，讓對方無法明確瞭解，違反誠信原則。[註9]

　　而且把定義跟義務混在一起，常常會讓定義沒辦法適當的發揮作用。舉例來說，如果把「通知」定義為「當事人依照本契約應該告知對方的事項」，可能反而忘記要約定清楚哪些事情，應該在什麼時間點以何種方式通知，以及實際上收受通知的人是誰？或是在定義條款中寫太多細節，結果跟後面實際條款內的約定事項發生衝突矛盾，究竟以定義條款為準，還是以後面的條款為準？也會發生問題。

(2) 不同字詞的定義不要混在一起

　　舉例而言：

例句八：

"Business Day" means the day that ABC Co. performs

9　以下各點注意事項參照 Stark 前揭著第 78-88 頁。

> ordinary operation（the "**Business**"）.

　　這句話會讓讀者很困擾。契約中的定義條款都是按照字母排列，讀者會從定義條款找「Business」的定義，可是「Business」卻被放在以「Business Day」為標題的定義中。

(3) 不要循環或冗贅的定義

　　循環定義就是被定義的字詞跟定義中的描述大致相同，並不是說被定義字詞絕對不能出現在定義的描述裡，而是說定義的描述中必須提供除了特定字詞以外，更多的訊息；例如「『商標』指的是一個或多個商標。」這種定義寫了跟沒寫差不了多少。出現循環定義有幾種處理方法，一種是還有其他說明得更清楚的方法，所以要再努力想一想怎麼說明，例如商標指的是哪一方當事人的商標，是否包括服務標章、公司名稱、網址？是否需要登記註冊？可以再多一些說明；另一種是真的沒什麼好多說的，字面意思就很清楚了，那就可以省些筆墨，直接刪除。

(4) 參照外部文件

　　契約的定義條款也可能會參照外部文件。例如當事人一方聲明並擔保「遵守相關法規」、「符合各種法規要求」。事實上，不太可能把法規內容全部抄進契約，但是，至少在定義中應盡可能描述重要的法規內容，否則這種包山包海的擔保最終只是空頭支票。

　　如果是參照其他相關契約，則可以用相關契約的前言加以特定，例如：

例句九：

"**Patent License Agreement**" means the Patent License Agreement, dated May 20, 2020, between ABC Co. and DEF. Inc., as amended from time to time.

被參照的相關契約可能會修訂，所以必須考慮清楚，如果這份契約的內容是要跟著相關契約最新的修訂內容，那就要加上「as amended from time to time」。相反地，如果是要鎖定相關契約最初的內容，那就改用「as of the date of its signing」。

(5) 善用定義條款

沒有必要定義的字詞就不要定義，一旦做了定義，就要發揮最大的功能，也就是精簡契約用字。比方說範例九的「Patent License Agreement」，既然已經加以定義，在契約各條中就一律使用「Patent License Agreement」，不需要再寫「Patent License Agreement, dated May 20, 2020, between ABC Co. and DEF. Inc., as amended from time to time」。

英文契約的核心
事項條款組合

英文契約的核心事項條款組合屬於起承轉合中的承。

我們稱為條款組合，是因為在契約中，通常這部分不只有單單一個條款，而是多個條款組成的。英文契約的核心事項條款組合大致由主要履約事項（Subject Matter Performance）、對價（Consideration）、契約期間（Term）、交割（Closing）等條款所組成。交割條款事實上也就是主要履約事項，只是交割需要安排特定日期特定地點由雙方彼此交付，而一般的主要履約事項與對價可能各付各的。並不是每一種契約都有交割條款，通常只有在證券或不動產交易中才會需要交割條款（註1）。另外，也有主張加上雙方履約內容的摘要。而這四類條款的排列順序，也是人言言殊。也有人把契約期間擺在最前面（註2）。並不是每一種交易都需要約定時間，只有持續的交易，才需要設定契約效期，對於這類交易，將效期編排在第二條，定義之後，主要履約事項之前，也是頗合乎邏輯的，而且可以避免遺漏。但是，也有很多人將效期跟終止事項放在同一條，而編排在更後面的條款。本章暫且按契約期間（Term）、主要履約事項（Subject Matter Performance）、對價（Consideration）、交割（Closing）與交割前後的義務的順序說明。

第一節　契約效期（Term）

並不是所有契約都需要約定效期，只有關於繼續進行交易的契約才需要效期，購買特定物的一次交易不需要效期，買一批貨交完貨、付完款，契約就結束了，不需要效期。

1　Tina L. Stark, *Drafting Contracts: How and Why Lawyers Do What They Do* 39-41（Wolters Kluwer, 2007）.

2　Sue Payne, *Basic Contract Drafting Assignments: A Narrative Approach* 373-382（Wolters Kluwer 2011）.

契約期間可以用宣示句的方式來寫，也就是以「契約期間」做主詞，「開始於特定時點」、「結束於特定時點」。開始與結束的動詞要用第三人稱單數，特定時點可以是某年某月某日，甚至指定到某時某分，也可以指定為某個事件的發生。

要留意指定的事件是否必然會發生，如果不一定會發生，那會讓整個契約的生效或失效跟某個條件聯結在一起。而如果負擔義務的一方可以控制讓那樣的事件發生或不發生，在美國，這樣的契約可能變成虛幻（illusionary）而不能執行，因為義務方可以控制自己受不受拘束的條件，他所負擔的義務就似有若無了。

除了某月某日、某個公眾節日，以及人的死亡之外，大概很少有什麼事件是必然會發生的。例如在Krell v. Henry（註3）這件案子中，英國國王愛德華七世（Edward VII）排定舉行加冕典禮，所以當事人才簽訂房屋租約，以便觀賞典禮。結果國王生病延後典禮的舉行，從而引發房子能不能退租的糾紛。當事人並未明確將加冕典禮的舉行列為契約條件或期限，但法院用履約目的受挫（frustration of purpose）替承租人找出一個默示的契約效力條件。契約如果約定以事件的發生做為起始日或終止日，必須周全考慮。

以下我們舉一糾紛實例說明契約效期條款寫作的問題：

案例一

Aliant Telecom v. Rogers（註4）
Rogers 公司向Aliant Telecom租用電線桿以鋪設線路提供電

3　[1903] 2 KB 740.
4　*Telecom Decision CRTC 2006-45*, http://www.crtc.gc.ca/eng/archive/2006/dt2006-45.htm, 最後點閱日：July 17，2023。

信服務，雙方簽訂支援架構協議（Support Structure Agreement, SSA）。SSA的期限條款如下：

"Subject to the termination provisions of [the SSA], [the SSA] shall be effective from the date it is made and shall continue in force for a period of five (5) years from the date it is made, and thereafter for successive five (5) year terms, unless and until terminated by one year prior notice in writing by either party.

（除適用SSA的終止條款之外，SSA應從簽約日起生效，並從簽約日起繼續在五(5)年內有效，而後再續約五年，除非任一方於一年之前以書面通知終止。）

簽約不久之後，Aliant Telecom的股東有變動，所以新的股東要求提前終止SSA，要將電線桿轉租給別人。於是Aliant Telecom就發通知終止SSA，但Rogers公司主張在第一個五年租期內還不能終止，要第二個租期才能終止。因為才剛鋪好線路沒多久，就要終止租約，承租人會嚴重虧損，所以雙方本來的意思是只有第二個租期之後才能終止。

審理本件糾紛的電信委員會認為按照文法來說，「除非任一方於一年之前以書面通知終止」這個句子之前被逗點分開，表示這句話既適用在第一個五年，也適用在第二個五年。

由上述案例，我們要強調的並不是逗點的文法規則，而是在契約造句上盡可能簡單、清楚、明瞭。句子寧短勿長。寧可多用句點，切分成多個句子，避免用太多逗點，將好多片段串成長句。太長、太複雜的句子難以理解，容易誤會。所以在效期條款也是一樣。最核心的句子就是這份契約從什麼時候開始生效，繼續多久有效。這句話說完後就用句點

結束。之後再處理自動續約或終止的問題。

此處要特別提醒。不續約與終止是不同的，不續約就寫「not to renew」，終止才寫「terminate」。不續約對應的是本來有自動續約的約定，但要排除自動續約。而終止是還在契約效期之內，但要提前結束。這兩者千萬不要搞混。前述案例二僅提到一年前終止，但卻未提到不續約，這是較特殊的地方，或許當事人在議約時沒有考慮清楚。一般來說，自動續約必定要加上可不自動續約的選項，通常是在原效期將要結束之前的一段時間，以書面通知他方不續約。如果沒有這樣的約定，可能被解釋為必定自動續約，沒有辦法不續約，那會沒完沒了，非常可怕。而且要特別強調，必須寫成，「除非一方通知他方不續約，否則自動續約」，不能寫成「除非雙方另有合意，否則自動續約」。因為後者也會被解釋成不續約還要雙方合意，若任一方不同意不續約，就自動續約，萬一是每一次都自動續約，那又是沒完沒了的契約，細思極恐。

如果約定除非事先通知，否則就自動續約，但是又希望調整續約條件，例如房東希望租約到期後自動續約，但是又希望漲租。最好約定清楚漲租幅度如何計算，如果只講租金再議，就容易發生糾紛。美國有實際的判決認定日後續約的租金如果約定再議，必須至少寫個算式，否則太過不明確，沒辦法發生拘束力[註5]。

提前終止契約的事項也可以在散場條款（Endgame Provisions）中加以處理。續約條款則通常放在契約期間條款中一併說明。

以下為契約期間條款範例：

5　*Joseph Martin, Jr., Delicatessen, Inc. v. Schumacher*, 417 N.E.2d 541, 543-544（N.Y. 1981）.

例句一：（註6）

Term. This Agreement's term begins on January 1, 20X5 and ends on December 31, 20X7. It automatically renews for successive one-year terms, unless either party exercises its option not to renew this Agreement. (The initial three-year period and each successive one-year renewal, a **"Term"**) To exercise its option not to renew at the end of a Term, a party must deliver a written notice to the other party that is received no later than thirty days before the last day of then existing Term.

第二節　主要履約事項（Subject Matter Performance）

　　主要履約事項就是這份契約的主要標的，依照作者的經驗，契約常見的主要標的大致分為五類，有一些較特殊少見的標的，或者仍然能套用到五種中的某一種，而無須增加類別，反而難以記憶。或者不在本書的探討範圍，可能涉及人身身份，而非商務契約。

　　五種標的就是人流、物流、金流、資訊流與商流。這是電子商務的流通概念，但也可借用作為契約標的的分類。契約的核心在於交換（bargain），而交換的標的不脫人、物、金、資訊與權利。人指的並不是販賣人口，而是針對人的服務進行交易，在法律上更細的分類則是僱傭、承攬與委任。物指各種有形體的物。金則指金錢。資訊亦容易理

6　改編自 Stark 前揭著第 104 頁。

解。而權利則指各種權利的交易。物是有形的，但是有形體的物之交流背後，往往另外涉及不同的權利流動，外觀上都是某甲交付一台機車給某乙，但背後的權利流動，可能是所有權的移轉，可能是某甲為了向某乙借錢，而將機車質押給某乙，也可能是某甲將機車出租給某乙，借此收費。所以同樣交付某物的行為，但會有不同的權利流動，與不同的法律關係。

標的意識至關重要，我們在第一章裡提到契約內容的四個重點，第一來者何人，第二有何指教，第三丁里相會終須一別，第四到哪裡輸贏。這第二點有何指教就是標的，契約就是要交換，所以要慎重思考交換的標的。常見一些契約寫了長篇累牘，卻看不出要交換何種標的，這樣的契約在法律上可能最後很難執行。記得以前在法學院上課，訴訟法的教授提到打官司訴之聲明很重要，就像去廟裡拜拜，到底要神明給什麼，求財、求姻緣、求健康，講得不清不楚是不行的，而訴之聲明就是要法院法官判命對造給你什麼，給錢、交付物品，做事、形成權利，或確認某種法律關係。簡單說就是給付、形成與確認。給付的標的，其實就是上述提過的人、金、物、資訊、權利。而以形成為標的寫入契約的情況或者極為少見，所以本書暫且不論，或者如同結婚、離婚、收養等人身關係，不在本書商務契約的主旨範圍內。至於確認，在英美法中很強調約因，就是對價關係，確認可能難以建構對價關係，已經有的關係再次確認，本身可能並非新的可用以交換的約因，也許確認的文件可以產生禁反言的效果，避免對方翻來覆去。但畢竟此種情況較少出現，所以本書也暫且不論。還有一些交易類型乍看之下似不屬於人、金、物、資訊、權利這五種標的，例如合夥、保證。那是因為這類契約涉及的標的較多，但歸根結柢，總是能找出人、金、物、資訊、權利。例如合夥，涉及的就是出錢，做事，劃權。例如保證，最終也是保證人要出錢。所以用人、金、物、資訊、權利作為契約標的之分類，大致上是可

行的。

以下我們逐一說明契約中的給付標的：

1. 物

作為契約標的之物，我們權且局限在有體物，因為無體物要劃歸為權利。有體物也就是動產與不動產等有形體的物。

2. 金

作為契約標的之金，就是指各種貨幣，是為了提高交換效率而運用的中介物。

3. 人

作為契約標的之人，指的是人的服務。依照臺灣民法債編的區分，人的服務大致有僱傭、承攬與委任三種原型。僱傭是買方花錢買下賣方的時間，要賣方做事，承攬是買方花錢要賣方做出某種成果，而委任則是買方花錢要賣方做具體而特定的事。僱傭強調的是時間，所以上班不能兼差，否則變成一物二賣就是違約了，但下班時間當然可以自行運用，所以也沒有責任制免費加班的道理。承攬的重點是要完成某種成果，例如訂製西裝，沒有交出西裝當然不能給錢。而委任的重點是事情要區分的很仔細，例如被告了，委託律師進行第一審的答辯，律師的報價是只針對一審當事人被告某幾個訴之聲明進行答辯，所以如果原告追加訴之聲明，可能就要追加報價，如果被告要反訴告回去，那也是另外

報價。一審判決出來而確定，律師的工作就結束了，要上訴也是另外報價。

4. 資訊

　　資訊包括各種各樣的資訊，例如技術訣竅（Knowhow），重點在於能說清楚明白就好。

5. 權利

　　權利，權利包括法律上明文設定的權利（物權），也包括一些法律上沒有設定的權利，比方說組織人事的決定權限，某一方可選派董事、總經理等職位，他方要在程序上相互配合。

　　主要履約事項基本上用給付句來寫作。當然，如果有本條款專用的字詞，就需要用宣示句對字詞做定義。履約過程中，如果需要他方對某些事項進行選擇加以特定，例如不同的包裝樣式，就用裁量選擇句來寫。給付句的寫法，就是以當事人一方為主詞，加上「shall」，以及動作、受詞、數量[註7]、履約時間、其他修飾（品質、規格、方法等），甚至條件。然而，國際貨物買賣，尤其大宗物品銷售，可能將主要履約事項的資料表格化，而不再是完整的句子。

　　每一種交易的給付內容有極大的差異，撰擬契約的人必須對交易結構與內容進行深入的研究。在契約寫作時，務必確認標的寫得一清二楚，沒有含混之處，否則容易引發爭議，以下我們用一個實例來說明：

7　在美國，標的數量比標的價格還重要，數量沒辦法確定，契約就沒有拘束力，價格不確定，法官或許還可以幫當事人事後定一個合理的價格。

案例二

Cantor Fitzgerald & Co. v. YES Bank Limited [註8]

Yes Bank 是印度一家銀行，財務狀況很糟。Yes Bank 為了解決財務問題，簽了一份委託契約，委託 Cantor Fitzgerald & Co. 協助募資。

契約約定委託事務「融資活動」（Financing）的範圍，以及 Cantor Fitzgerald & Co 可以從 Financing 得到的資金中收取2%作為佣金。

Cantor Fitzgerald & Co 搞了一陣子，但沒有實質結果。之後印度央行介入，由官股銀行收購 Yes Bank，然後再對外公開發行新股（FPO）募資。雖然這個過程主要是由印度央行主導，但是 Cantor Fitzgerald & Co 主張有權從公開發行新股得到的資金中收2%。

兩造爭議關鍵點在於契約中對於服務內容 Financing 的定義。契約文字是這麼寫的："the private placement, offering or other sale of equity instruments in any form"

Yes Bank 主張，上面這句話要讀成 private placement, private offering or other private sale of equity instrument，所以做FPO是公開發行，不是 private，Cantor Fitzgerald & Co 不能收2%。

Cantor Fitzgerald & Co 提出相反的主張，private 只是修飾 placement，不修飾 offering or other sale of equity instruments in any form，所以 offering or other sale of equity instruments in any form 也包括 FPO 公開發行。

8　[2023] EWHC 745（Comm），https://www.bailii.org/ew/cases/EWHC/Comm/2023/745.html，最後點閱日：July 16，2023。

　　法官先用文法討論，說一個形容詞後面加好幾個名詞，原則上，形容詞應該解釋成修飾後面的每一個名詞。

　　也有論者反對案例一法官的文法見解[註9]。如果就證券財顧實務來說，placement跟offering基本上是一樣的意思，就是發行新股，反而sale of equity instruments通常是講老股的買賣。placement跟offering的差別只在於我們常講private placement（私募），但卻說public offering（公開發行），而較少講private offering或public placement。所以公開發行新股來募資，是用offering，但發行新股洽特定人認購，卻是用placement。而sale of equity instruments需要財顧協助的通常是私下交易，因為公開的意思是要到證券交易所裡對不特定人交易，但財務顧問沒辦法插手對不特定人的交易，只能協助非公開已發行股份的私人間交易，或者已公開發行股份巨額交易、配對交易，都必須鎖定對象。所以用證券財顧實務來說，offering是公開的，但sale of equity instruments卻是私下的。

　　本書的重點不在個案糾紛的解決，只是說明在契約中寫給付標的時務必留意一定要一清二楚，有任何疑慮可能性時，可以用表列（tabulation）的方式，分點分段寫下，例如(1)private placement, (2) public offering (3)other private sale of equity instruments in any form。

　　以下提供二個樣本說明主要履約事項的構成句型與外觀：

9　https://www.adamsdrafting.com/cantor-fitzgerald-co-v-yes-bank-limited-syntactic-ambiguity/，最後點閱日：July 16，2023。

例句二：（註10）

Agency Contract

3. The Agent shall use best effort to (i) obtain business for the Principal; (ii) serve the Principal's interests; (iii) provide all information necessary in promoting business; and (iv) especially inform the Principal immediately about every order received. The Agent may not deviate from the prices, delivery and payment conditions of the Principal without his consent.

例句三：（註11）

Cross-license between a foundry and a customer for joint product design and development of products marketed by each party

The OWNERSHIP and LICENSING SECTION. X grants a license to Y for making integrated circuits and related software LICENSED PRODUCTS. Y grants to X a license to make or have Third Parties make solely for X LICENSED PRODUCTS. An existing product of Y is excluded from LICENSEDP RODUCTS, but updates to this product which incorporates X 's technology are LICENSED PRODUCTS for royalty purposes. The license to

---

10 改編 International Union of Commercial Agents and Brokers, *Agency Contract* (2001) 轉引自 Juris International, http://www.jurisint.org/en/con/341.html, latest accessed on 2011/6/3.

11 改編 Licensing Executive Society International, *Cross-license between a foundry and a customer for joint product design and development of products marketed by each party* (2001)，轉引自 Juris International, http://www.jurisint.org/en/con/361.html, latest accessed on 2011/6/3.

Y for a LICENSED PRODUCT is exclusive for time, extendable as X's option for an additional year of exclusivity. The license granted to Y for UPDATED LICENSED PRODUCTS is exclusive for_____, which Y have the option extend for an additional year of exclusivity.

第三節　對價（Consideration）

對價可能是金錢，也可能不是金錢。總之，對價條款是要交代另一方當事人必須給付的內容。其句型架構跟主要履約事項一樣，以當事人一方為主詞，加上「shall」，以及動作、受詞、數量[註12]、履約時間、其他修飾（品質、規格、方法等），甚至條件。

如果對價是金錢，還必須考量計價方式、各種支付工具如何指定、付款期限、收款人帳戶、信用風險、費用、幣別、匯率、利息、衍生費用、擔保、帳冊保管與核對等問題。[註13]此處不擬深入討論。以下提供樣本例示。[註14]

例句四：

4. PRICE_____

　　Price *(specify whether total price and/or price per unit of measu-*

12　在美國，標的數量比標的價格還重要，數量沒辦法確定，契約就沒有拘束力，價格不確定，法官還可以幫當事人事後定一個合理的價格。

13　參閱 Stark 前揭著第 97-102 頁。

14　International Trade Centre UNCTAD/WTO, *International Commercial Sale of Perishable Goods: Model Contract and Users's Guide*（1999）轉引自 Juris International, http://www.jurisint.org/en/con/339.html, latest accessed on 2011/6/3.

rement, specify the amount in both figures and words, and the currency)

_____

_____

Method for determining the price *(if appropriate)*_____

_____

_____

Where the price has not been and Call1noblt e determined, it shall be thCltg enerally charged, in the same trade, for such goods delivered under comparable circumstances or, if such price cannot be established, a reasonable price.

5. PAYMENT_____

Means of payment *(e.g. cash, cheque, bank draft, transfer)*

_____

Details of Seller's bank account *(if appropriate)*_____

_____

Unless otherwise agreed, the amounts due shall be transferred by teletransmission to the Seller's bank account, and the Buyer shall be deemed to have performed its payment obligations when the said amounts have been received by the Seller's bank._____

Payment of the price shall be made within 30 days after the date of invoice, unless the parties agree a different period hereafter:

THE PARTIES MAY CHOOSE A PAYMENT ARRANGEMENT AMONG THE POSSIBILITIES SET OUT BELOW I IN WHICH CASE THEY SHOULD SPECIFY THE ARRANGEMENT CHOSEN AND PROVIDE THE CORRESPONDING

DETAIlS:

☐ PAYMENT IN ADVANCE

Amount due *(i.e. all or part of the price, or expressed as a percentage of the totol price)*

Latest date for payment to be received by Seller's bank

_____

Special conditions applying to this payment *(if any)*

_____

In the event that the advance payment does not correspond to the total price, the balance due shall be payable within 30 days of the date of invoice, in accordance with the conditions set out above, *unless otherwise specified hereafter:*_____

_____

_____

☐ PAYMENTB Y DOCUMENTARCYO IIECTION

Amount to be paid *(specify whether total price or price per delivery instalment)*

Latest date for payment_____

Means of payment: DIP *(i.e. documents against payment),* unless the parties specify D/ A *(i.e. documents against acceptance)* hereafter:

Payment by documentary collection shall be subject to the ICC Uniform Rules for Collections.

The documents to be presented are specified at Article 6 below.

☐ PAYMENT BY IRREVOCABLED OCUMENTARYC REDIT

The Buyer must arrange for an irrevocable documentary

credit in favour of the Seller to be issued by a reputable bank, subject to the ICC Uniform Customs and Practice for Documentary Credits. The issue must be notified at least 14 days before the agreed date for delivery, or before the beginning of the agreed delivery period specified at Article 3 above, as appropriate, *unless the parties agree otherwise as specified hereafter*. Latest agreed date for issue:_____

The credit shall expire 14 days after the end of the period or date of delivery specified in Article 3 above, *unless otherwise agreed hereafter:*_____

The documentary credit does not have to be confirmed, *unless the parties agree otherwise, as specified hereafter*:_____

All costs incurred in relation to confirmation shall be borne by the Seller, *unless the parties specify otherwise hereafter:* _____Unless otherwise agreed, the documentary credit shall be payable at sight and allow partial shipments and transhipments.

☐ PAYMENT BACKED BY BANK GUARANTEE

The Buyer shall provide, at least 30 days before the agreed date of delivery or the beginning of the agreed delivery period specified at Article 3 above as appropriate, unless the parties specify hereafter some other date:_____, either a first demand bank guarantee subject to the ICC Uniform Rules for Demand Guarantees, or a standby letter of credit subject either to such Rules or to the ICC Uniform Customs and Practice for Documentary Credits, in either case issued by a reputable bank.

☐ OTHER PAYMENT ARRANGEMENTS

第四節　交割（Closing）與交割前提條件與交割前後之義務

交割或過戶適用在需要以特定手續辦理完成產權移轉的情況。例如不動產我們常稱為過戶，而證券所有權移轉，我們則常稱為交割。如果涉及這類權利移轉，通常會約定時間地點，來辦理交割或過戶。交割過戶條款務必要列明應交付的文件、應簽署的文件，以及應向政府機關申報的事項。另外也可能涉及各種稅賦費用負擔期間的切分。

在雙方談妥交易的時候，交易的條件可能還沒完全做到，例如買房子，賣方答應將外牆重新粉刷成買方指定的顏色，這時就需要給賣方時間以完成這項條件，或者買方貸款還沒辦下來，但是雙方相談甚歡，所以買方繳一筆定金，繼續申辦貸款，貸款辦下來了再交割。這些事項可能列為交割前義務，或者設定為交割的前提條件，或是整個契約生效或失效的條件，之後若沒有完成這些待辦事項，各方當事人就可能有暫停履約、沒收定金、解除契約、回復原狀等回應。設定成交割前義務，跟設定成條件，效力是不同的。一方若違反，就有違約的責任，可能要賠償，而他方並不因此就不用履行對待給付義務，但設定成條件，或許只是契約不生效，或者他方可以解除契約，或者他方暫時不負擔對待給付的義務。

另外，不動產或證券移轉、甚至整間公司的移轉，必然有相當複雜

的手續，在執行過程中難免疏漏，所以有時也會加上交割後的協力義務，要求賣方配合買方辦理後續程序。

交割安排條款範例如下：

例句五：（註15）

6.1 Closing. The consummation of the transation this Agreement contemplates (the "Closing") will take place at the offices of Wnag & Kao LLP, 1000 Tibet Street, Kaohsiung, Taiwan, 804 on May 20, 2016 at 9:30 am local time or on another date and time both parties agree and no later than June 15, 2016 (the date and time of the Closing, the "Closing Date")（註16）.

6.2 Closing Deliveries.

(a) **Seller's Deliveries.** At the Closing, the Seller shall deliver to the Buyer the following:

(i) A bill of sale for the Purchase Assets.

(ii) An assignment of each real property lease under which the Seller is lessee, *in form satisfactory to the Buyer.*

(iii) Assignments for all funds on deposit with banks or other Persons which are Purchased Assets, *in form reasonably satisfactory to the Buyer.*

(b) **Buyer's Deliveries.** At the Closing, the Buyer shall

---

15 Stark 前揭著第105-106頁。

16 如果在契約中不只一個地方提到"Closing"以及"Closing Date"，也可以把嵌入式定義改成獨立式的定義，放在定義條款中。

deliver to the Buyer the following:
    (i)   xxxxx
    (ii)

　　以下用例句六說明併購契約中設定標的公司與子公司在交割之前應該做哪些事、不應該做哪些事的規定，包括要維持公司的正常營運、不可以修章、不可以發行新股、不可以處分標的公司或子公司的重要資產、不可以配股配息、不可以合併分割股權或做其他調整、不可以設立子公司、分公司或關閉子公司、分公司等非常瑣細而詳盡的規定。在設定這類作為義務與不作為義務時，需要審慎考慮如何訂定明確、可行而實際的義務。禁止的行為常較容易明確列舉，不可以做這個，不可以做那個，像以下例句六的b.列了26項禁止規定。但是要求做什麼事的約定，有些還是明確，比如該報稅繳稅、有稅務機關的處分或訴訟等事件要立即通報，這些很清楚，對帳單就好，像以下例句六的c。但其他的可能就不是那麼有效了，像是例句六的a要求維持正常營運、盡商業上努力維持商譽價值，這些要求與其設定成義務，不如透過其他合理客觀的機制來落實，比如說在接近交割的時點再重新鑑定商譽價值，讓買價跟交割前的商譽價值連動，或者讓買價跟交割前的營收與利潤連動。

例句六：

ARTICLE V CONDUCT OF BUSINESS PENDING THE MERGER [註17]
    SECTION 5.01　　Conduct of Business by the Company

17　ATHENAHEALTH, INC.,Echo Merger Sub, Inc. Epocrates, Inc. 合併契約，https://www.sec.gov/Archives/edgar/data/1131096/000113109613000008/mergeragreement.htm，最後點閱日：July 23，2023。

Pending the Merger.

a. The Company covenants and agrees that, between the date of this Agreement and the Effective Time, except as set forth in Section 5.01 of the Company Disclosure Schedule or as expressly contemplated by any other provision of this Agreement, unless Parent shall otherwise provide prior consent in writing, which may not be unreasonably withheld, conditioned or delayed, provided, however, that the Company shall not be required to obtain consent if doing so may violate antitrust Law (provided that the Company provides notice of such belief to Parent promptly):

   i.   the businesses of the Company and the Subsidiaries shall be conducted only in, and the Company and the Subsidiaries shall not take any action except in, the ordinary course of business and in a manner consistent with past practice; and

   ii. the Company and each of its Subsidiaries shall use their commercially reasonable efforts to preserve substantially intact the business organization goodwill and other existing assets of the Company and the Subsidiaries, to keep available the services of the current officers, employees and consultants of the Company and the Subsidiaries, to maintain and preserve intact the current relationships of the Company and the Subsidiaries with customers, suppliers, distributors, creditors and other persons with which the Company or any Subsidiary has significant business relations, and to comply in all material respects with applicable Law.

b. By way of amplification and not limitation, except as expressly contemplated by any other provision of this Agreement or as set forth in Section 5.01 of the Company Disclosure Schedule, neither the Company nor any Subsidiary shall, between the date of this Agreement and the Effective Time, directly or indirectly, do, or propose to do, any of the following without the prior written consent of Parent, which may not be unreasonably withheld, conditioned or delayed, provided, however, that the Company shall not be required to obtain consent if doing so may violate antitrust Law (provided that the Company provides notice of such belief to Parent promptly):

    i. amend or otherwise change its Certificate of Incorporation or Bylaws or equivalent organizational documents;

    ii. issue, sell, pledge, dispose of, grant, award or encumber, or authorize the issuance, sale, pledge, disposition, grant, award or encumbrance of, any shares of any class of capital stock of the Company or any Subsidiary, or any options, warrants, convertible securities or other rights or equity of any kind to acquire any shares of such capital stock, or any other ownership interest (including, without limitation, any phantom interest), of the Company or any Subsidiary (except for the issuance of Shares issuable pursuant to employee stock options outstanding on the date hereof);

iii. sell, pledge, dispose of, grant or encumber, or permit an encumbrance to exist on, or authorize the issuance, sale, pledge, disposition, grant or encumbrance of any material assets of the Company or any of its Subsidiaries;

iv. declare, set aside, make or pay any dividend or other distribution, payable in cash, stock, property or otherwise, with respect to any of its capital stock;

v. adjust, reclassify, combine, split, subdivide or redeem, or purchase or otherwise acquire, directly or indirectly, any of its capital stock;

vi. (A) acquire (including, without limitation, by merger, consolidation, or acquisition of stock or assets or any other business combination) any corporation, partnership, other business organization or any division thereof or any material amount of assets; (B) incur any indebtedness for borrowed money or issue any debt securities or assume, guarantee or endorse, or otherwise become responsible for, the obligations of any person, or make any loans or advances, or grant any security interest in any of its assets except in the ordinary course of business and consistent with past practice; (C) enter into any contract or agreement other than in the ordinary course of business and consistent with past practice and provided such contract or agreement shall not involve any discounts in excess of

25%; (D) authorize, or make any commitment with respect to, any single capital expenditure which is in excess of $100,000 or capital expenditures which are, in the aggregate, in excess of $450,000 for the Company and the Subsidiaries taken as a whole, except as consistent with Company business plans that have been disclosed to Parent; or (E) enter into or amend any contract, agreement, commitment or arrangement with respect to any Contract or matter set forth in this Section 5.01(b)(vi);

vii. (A) hire any employees in the position of Vice President or above; (B) take any action which could reasonably be expected to give rise to a claim for "good reason" or any similar claim by any employee; or (C) terminate the employment of employees in the position of Vice President or above (other than for cause, so long as prior notice of such termination is given to Parent);

viii. hire any additional employees or increase the compensation payable or to become payable or the benefits provided to its directors, officers, employees or independent contractors, or grant any severance or termination pay to, or enter into any employment or severance agreement with, any director, officer employee or other independent contractors of the Company or of any Subsidiary, or establish, adopt, enter into or amend any collective bargaining, bonus, profit-sharing, thrift, compensation, stock option, restricted stock, pension, retirement, deferred compensation, employment, termination, severance or other plan, agreement, trust, fund, policy or arrangement for the benefit of any director, officer, employee or independent contractor, or loan or advance any money or property to any director, officer, employee or independent contractor, or grant any equity or equity-based awards;

ix. exercise its discretion with respect to or otherwise voluntarily accelerate the lapse of restriction or vesting of any Company Stock Option or Company RSU as a result of the Merger, any other change of control of the Company (as defined in the Company Stock Option Plans) or otherwise;

x. terminate, discontinue, close or dispose of any plant, facility or other business operation, or lay off any employees (other than layoffs of less than 10 employees in any six-month period in the ordinary course of business consistent with past practice) or implement any early retirement or separation program, or any program providing early retirement window benefits or announce or plan any such action or program for the future;

xi. take any action, other than reasonable and usual actions in the ordinary course of business and consistent with past practice, with respect to accounting policies or procedures;

xii. make, change or rescind any material Tax election,

settle or compromise any material United States federal, state, local or non-United States income tax liability, change an annual accounting period, adopt or change any accounting method, file any material amended Tax Return, enter into any closing agreement, surrender any right to claim a refund of material Taxes, or consent to any extension or waiver of the limitation period applicable to any Tax claim or assessment relating to the Company or any of its Subsidiaries;

xiii. pay, discharge or satisfy any claim, liability or obligation (absolute, accrued, asserted or unasserted, contingent or otherwise), other than the payment, discharge or satisfaction, in the ordinary course of business and consistent with past practice, of liabilities reflected or reserved against in the 2012 Balance Sheet or subsequently incurred in the ordinary course of business and consistent with past practice;

xiv. enter into, amend, modify, in a manner that is adverse to the Company, or consent to the termination of any Material Contract (other than terminations consented to in the ordinary course of business, consistent with past practice and without financial penalty to the Company or any Subsidiaries), or enter into, amend, waive, modify, in a manner that is adverse to the Company, or consent to the termination of the Company's or any Subsidiary's rights thereunder (other than terminations consented to in the ordinary course of business, consistent with past practice and without financial penalty to the Company or any Subsidiaries);

xv. commence, compromise, settle or come to an arrangement regarding, or agree or consent to compromise, settle or come to an arrangement regarding, any Action;

xvi. permit any item of Company Owned Intellectual Property to lapse or to be abandoned, dedicated, or disclaimed, fail to perform or make any applicable filings, recordings or other similar actions or filings, or fail to pay all required fees and taxes required or advisable to maintain and protect its interest in each and every item of Company Owned Intellectual Property;

xvii. (A) abandon, disclaim, dedicate to the public, sell, assign or grant any security interest in, to or under any Company Owned Intellectual Property or Company Licensed Intellectual Property, including failing to perform or cause to be performed all applicable filings, recordings and other acts, or to pay or cause to be paid all required fees and material Taxes, to maintain and protect its interest in Company Owned Intellectual Property or Company Licensed Intellectual Property; (B) grant to any third party any license, or enter into any covenant not to sue, with respect to any Company Owned Intellectual Property or Company Licensed Intellectual Property, except in the ordinary course of business consistent with past practice; (C) develop, create or invent any Intellectual Property jointly with any third party, except under existing arrangements that have been disclosed to Parent; (D) disclose or allow to be disclosed any confidential information or confidential Company Owned Intellectual Property or Company Licensed Intellectual Property to any person, other than employees of the Company or its Subsidiaries that are subject to a confidentiality or non-disclosure covenant protecting against further disclosure thereof, except under existing arrangements that have been disclosed to Parent; or (E) fail to notify Parent promptly of any infringement, misappropriation or other violation of or conflict with any material Company Owned Intellectual Property or Company Licensed Intellectual Property of which the Company or any of its Subsidiaries becomes aware and to consult with Parent regarding the actions (if any) to take to protect such Company Owned Intellectual Property or Company Licensed Intellectual Property;

xviii. fail to make in a timely manner any filings with the SEC required under the Securities Act or the Exchange Act or the rules and regulations promulgated thereunder;

xix. fail to maintain (with insurance companies substantially as financially responsible as its existing insurers) insurance in at least such amounts and against at least such risks and losses as are consistent in all material respects with the Company's and its Subsidiaries' past practice;

xx. write down or write up or fail to write down or write up the value of any receivables or revalue any assets of the Company, other than in the ordinary course of business and in accordance with GAAP;

xxi. form any Subsidiary;

xxii. enter into any agreement, understanding or arrangement with respect to the voting or registration of the capital stock of the Company or any of its Subsidiaries;

xxiii. take any action to render inapplicable, or to exempt any third party from, the provisions of any antitakeover Laws of any Governmental Authority;

xxiv. knowingly take or fail to take any action in breach of this Agreement or for the purpose of materially delaying or preventing (or which would be reasonably expected to materially delay or prevent) the consummation of the transactions contemplated hereby;

xxv. authorize, recommend, propose or announce an intention to adopt a plan of complete or partial liquidation or dissolution of the Company or any of its Subsidiaries; or

xxvi. announce an intention, authorize, resolve, enter into any formal or informal agreement or otherwise make a commitment, to do any of the foregoing or any other action inconsistent with the foregoing.

c. In addition, between the date of this Agreement and the

Effective Time, the Company and its Subsidiaries shall

(i) prepare and timely file all material Tax returns required to be filed (and shall provide Parent with copies of such material Tax returns a reasonable period of time prior to filing for Parent's review, but not approval, and shall incorporate in any final draft Parent's reasonable comments, to the extent consistent with applicable law),

(ii) timely pay all Taxes shown to be due and payable on such material Tax Returns, and

(iii) promptly notify Parent of any notice of any suit, claim, action, investigation, audit or proceeding in respect of any Tax matters (or any significant developments with respect to ongoing suits, claims, actions, investigations, audits or proceedings in respect of such material Tax matters).

附隨義務條款組合
與保密條款

　　在核心事項條款組合，也就是主要給付義務條款組合之後，接續的就是附隨義務之約定，附隨義務條款組合通常包括聲明與擔保、契約效力條件、保密義務、競業禁止、獨家交易、廉潔義務、彌償義務等條款。並不是每一份契約都會有上述這些條款，仍是要看個別交易是否需要這些約定。

　　聲明與擔保是英美契約法上的概念，其功能在於確認當事人之所以願進行交易的基礎事實真的存在。聲明是聲明過去與現在的事實，擔保可以擔保過去、現在與未來的事實。聲明與擔保的寫法，我們在本書第二章裡已經說明，所以我們此處不再重複。每份契約要聲明擔保的內容各有不同，例如僱傭契約可能需要受僱傭者聲明具有特定的資格條件，有相關的學歷、經歷，而且已經從前一個工作離職，並且沒有競業限制使其不能在新公司任職，也要聲明不會用前一個僱主的機密資訊、或智慧財產來執行新公司的工作等等。而買賣契約則需要加上瑕疵擔保與保固擔保之內容。

　　如果是較複雜的契約，或標的金額龐大的契約，通常會特別擬訂一條，詳列需要符合什麼樣的條件，才會繼續進行契約，或者若發生什麼樣的條件，契約就不進行了，例如併購契約、不動產買賣契約。至於條件的寫法，我們在第四章也說明過，此處不贅述。所以我們以下將在本章說明保密條款，並在後續幾章說明競業禁止、獨家交易、廉潔義務、彌償義務等條款。

　　當交易內容涉及敏感資訊時，當事人可能希望納入保密條款。保密條款的內容包括機密資訊的提供、機密資訊的定義、機密資訊的使用限制、保密方法與標準、保密期限；例外許可揭露或使用事由（例如政府機關強制揭露）、政府機關強制揭露時的因應、機密資訊的所有權；以及機密資料的銷毀或返還、違約救濟。

第一節　機密資訊的提供

之所以要簽訂保密條款，是因為締約當事人間有機密資訊的流通，此種流通可能是單向的。從而只有一方當事人負擔保密義務，也可以是相互的，也就是各方當事人均互負保密義務。

若是單向的機密資訊提供，條款內當事人的稱謂就可以延用契約最開頭部分的稱謂。若是雙向提供，條款內當事人的稱謂就必須另行界定，例如揭露方（Discloser）與收受方（Recipient）。

機密資訊的揭露有其時間、有其目的，藉著交代時間與目的，可以限定被保護的機密資訊範圍，與將來收受資訊一方的使用範圍。若保密條款欠缺這些相關訊息，可能不夠明確，一方認定保密的範圍較寬，而另一方認定保密的範圍較窄，日後便容易引發爭執。

例句一：

Either party（the "Discloser"）may provide the other party（the "Recipient"）Confidential Information（as defined in subsection x of this Article）MMDDYY through MMDDYY for the purposes of＿＿＿＿（the "Subject Matter"）.

第二節　機密資訊的定義

機密資訊的有四種定義方法：(1)抽象定義加上舉例、(2)標示或書面指明、(3)混合式、(4)負面排除，說明如下：

1. 抽象定義加上舉例

如果坐著想像如何對機密資訊做出抽象定義，可能想上大半天也想不出來，幸好，已有許多法律文獻提供我們參考資訊，例如可參考《與貿易有關的智慧財產權協定》（Agreement on Trade-Related Aspects of Intellectual Property Rights, TRIPS）與各國的營業秘密法規。

TRIPS第39條第1項規定要求簽署協定的各會員國保護未揭露的資訊，而同條第2項對未揭露資訊定義如下：

"Natural and legal persons shall have the possibility of preventing information lawfully within their control from being disclosed to, acquired by, or used by others without their consent in a manner contrary to honest commercial practices so long as such information:

(a) is secret in the sense that it is not, as a body or in the precise configuration and assembly of its components, generally known among or readily accessible to persons within the circles that normally deal with the kind of information in question;

(b) has commercial value because it is secret; and

(c) has been subject to reasonable steps under the circumstances, by the person lawfully in control of the information, to keep it secret."

因此，不論是TRIPS簽署國，或非簽署國，在制定營業秘密保護相關法律時，都可能參考TRIPS對未揭露資訊的定義：(1)資訊整體或其各部分之設定，與組成非屬通常處理該類資訊的人普遍知道或可輕易確

認的；(2)因為隱密而具有商業價值；而且(3)合法擁有資訊者已採取個別情況下合理的保密措施。

我們接著檢視臺灣的營業秘密法第二條：

「本法所稱營業秘密，係指方法、技術、製程、配方、程式、設計或其他可用於生產、銷售或經營之資訊，而符合左列要件者：

一、非一般涉及該類資訊之人所知者。

二、因其秘密性而具有實際或潛在之經濟價值者。

三、所有人已採取合理之保密措施者。」

上述法條的定義包括抽象與舉例，我們首先必須識別哪些文字是抽象定義，哪些文字是舉例。舉例的部分就是「方法、技術、製程、配方、程式、設計」。而抽象定義的部份，如果是交集（AND），定義項目愈多，符合定義的對象範圍愈小，就像切蛋糕那樣，切愈多刀，蛋糕愈小。如果是聯集（OR），定義項目愈多，則範圍愈大。臺灣營業秘密法第2條對營業秘密的抽象定義是交集，而且有四個抽象要件，前三個要件顯而易見，就是「非一般涉及該類資訊之人所知」、「因其秘密性而具有實際或潛在之經濟價值」、與「所有人已採取合理之保密措施。」至於第四個要件則是藏在舉例的後面「可用於生產、銷售或經營之資訊」。方法、技術、製程、配方、程式、設計都是舉例，但法條的意思，並不是所有方法、技術、製程、配方、程式、設計都算得上營業秘密，必須是符合前述四個要件的，才算是營業秘密。我們以下來看一個爭議實例：

案例一 [1]

被告王某任職於A房仲公司，並與A公司簽訂資訊使用承諾書），對A公司內部未對外公開之業務資訊有保密之義務，亦為受公司委任處理財產事務之人。新北當時有戶凶宅，屋主是透過A公司另一個房仲買下該棟房屋，買的時候，A公司說不是凶宅，後來屋主要轉賣這間房屋，被告王某受另一家房仲業者的營業員（代表買方）請託，登入公司電腦系統查詢凶宅資料，確實是凶宅，用手機翻拍，發送給請託人，請託人再提供給買方。結果交易當然吹掉了，A公司被凶宅屋主求償，所以公司對被告提告，主張被告違反營業秘密法。

法官認為房屋是不是凶宅，除了屋主知道之外，周遭鄰居、當地村里長、警察機關等亦應有所耳聞，所以凶宅資訊是否符合營業秘密法第二條「非一般涉及該類資訊之人所知」要件，不無疑問。而且「凶宅」、「海砂屋」等，社會大眾對之大多存有嫌惡畏懼之心理，造成該等建物之市場接受程度及交易價格低落，故房屋為凶宅或海砂屋之資訊，在房地產交易市場及實務經驗中，並無加以利用而獲利之可能。所以也不符合營業秘密法第二條「二、因其秘密性而具有實際或潛在之經濟價值」。

也有學者 [2] 認為凶宅資訊省去購屋者實際調查的成本，也能讓房仲避免介紹凶宅而被房屋買方追訴求償，降低營業風險，而應屬於有經濟價值的營業秘密。在前述案子中，若認定營業員王某違反營業秘密

1　臺灣高等法院104年度上訴字第339號刑事判決。
2　國立政治大學科技管理與智慧財產研究所副教授陳秉訓撰文，「凶宅資訊」是不是營業秘密法所保護的標的？北美智權報第226期，http://www.naipo.com/Portals/1/web_tw/Knowledge_Center/Infringement_Case/IPNC_181213_0501.htm

法，效果上可能讓人誤會隱匿凶宅資訊才是正當的，而且A公司被求償，主要是因為之前沒查清楚就跟屋主說不是凶宅，屋主才買下來，而並不是往後來把凶宅資訊，間接透露給新的買方。

提出這個實例是要讓讀者瞭解，有些資訊本身未必具有經濟價值，未必能直接拿來用於生產、銷售或經營，例如凶宅資訊，或者像3M的便利貼，據說是研發失敗的膠水，研發失敗的東西本來不能用來生產、銷售或經營。但是一則可以避免重蹈覆轍，二則換個方向可能搞出意想不到的商機，所以有沒有價值往往在轉念之間。營業秘密法是刑事法律，對於保護的對象可能設定較嚴苛的要件，避免人們誤觸法網，動輒得咎，監獄大爆滿。但契約中的保密條款未必要那麼嚴苛。

我們再來看看美國的統一營業秘密法（Uniform Trade Secrets Act）第1條第1項第4款如何定義Trade Secret：

"'Trade secret' means information, including a formula, pattern, compilation, program, device, method, technique, or process, that: (i) derives independent economic value, actual or potential, from not being generally known to, and not being readily ascertainable by proper means by, other persons who can obtain economic value from its disclosure or use, and (ii) is the subject of efforts that are reasonable under the circumstances to maintain its secrecy."

美國統一營業秘密法第1條第1項第4款對Trade Secret的定義同樣包括抽象定義與舉例，舉例的部分就是配方、模式、組合、程式、裝置、方法、技巧或程序這類資訊，但必須符合以下要件：(i)因為該資訊揭露或運用而可能獲益的人，一般人不知道，且無法用適當的方法輕易確認、有獨立的經濟價值，無論是實際的價值或潛在價值，再而(ii)已採取個案情況下的合理努力加以保密。

　　所以與臺灣的定義相比，美國的營業秘密定義多了一項要件，也少了兩項要件，還有一項細微的差異。多的部分是業內人士無法輕易查知，例如把機器拆開來就知道怎麼做的，或者嚐一口就知道放了哪些材料加以瞭解，這部分可能影響不大，畢竟如果是業內人士雖不確知，但只要看到外觀就能明瞭內涵的，也難以認定屬於機密。少的部分是「可用於生產、銷售或經營」，以及美國的法條定義要求「獨立的」經濟價值，臺灣的法律條文雖然並未明定經濟價值必須是獨立的，但實務裁判上，仍加上獨立的要件所以如果A資訊本身沒有價值，但合併B資訊後具有價值，A資訊仍然不是營業秘密。還有一項細微的差異在於臺灣的定義是「非『一般涉及該類資訊之人』所知者」而美國則是「『因為該資訊揭露或運用而可能獲益』的人一般不知道」。

　　中國大陸的不正當競爭法第9條第4項規定「本條所稱的商業秘密，是指不為公　所知悉，具有商業價值並經權利人採取相應保密措施的技術信息、經營信息等商業信息。」與臺灣營業秘密法相較，「不為公眾所知悉」跟「非一般涉及該類資訊之人所知者」不同，前者範圍可能較大，後者範圍可能較小，有些資訊一般人不知道，但業內人士知道。臺灣的定義包含「可用於生產、銷售或經營之資訊」，而大陸的則是「技術信息、經營信息等商業信息」。

　　上面討論的課題是國際條約與各國法律對機密資訊的保護，各國法律通常將受保護的機密資訊稱為營業秘密（trade secret），在臺灣，刑法本來也有妨害秘密罪章，其中包括非法手段探知資訊、專業人士洩露客戶當事人資訊，以及依法或依契約有守密義務者，不正當洩露。另外也有個人資料保護法的規定。但各國這些法律多半是出於公益保護，保護的方法可能是用刑責威嚇不法行為，並不涉及私人契約，國家已經對於保護的資訊對象做了確切的定義。但我們在簽訂契約，如果不考慮刑

事責任問題，單純用民事契約來保護機密資訊，我們可以對機密資訊的定義方法，可以與國家法律出於公益保護而定義的內容不同。

但是私人契約的定義不能悖離受保護資訊的核心要件，亦即(1)業內人士不知或無法輕易查知，以及(2)資訊擁有者已採取合理保密措施。在上述的核心要件之外，我們可以增加一些聯集或交集要件來放大或縮小保護範圍。增加的要件項目可能包括有無商業價值，這點可以分成更細膩的情況，其一是沒有商業價值仍受保護（例如敏感性的資訊）；其二是資訊本身沒有獨立的商業價值，但合併其他資訊後可能有商業價值，仍予保護；其三是必須有獨立的商業價值才受保護。而商業價值還可以再細分成實際的商業價值，或潛在的商業價值。另外也可以針對是否能用於生產、經營、銷售做出規範。我們需要特別留意，在契約中，(1)明寫特定要件，(2)明文排除特定要件，跟(3)不寫特定要件，三種做可能有不同的效果。舉例來說(1)本契約所稱之機密資訊必須具有商業價值，(2)本契約所稱之機密資訊只要符合上述要件，無論是否具商業價值均可，以及(3)不提商業價值。當然是以(1)或(2)最直接明白，(3)的做法可能還是會引起糾紛，但契約談判有時僵持不下，採取模糊的對應方法也是不得已的。

2. 在媒體上標明機密屬性

由於抽象定義加上舉例的方法在個案判斷上仍可能產生究竟是不是機密資訊的爭執，因此，更為簡便的作法則是由資訊提供人自己標示或告知為機密，若是用有形的媒介（例如紙張、儲存設備，檔案）提供資料，當然可標示為機密，若是沒有具體形式的媒介，例如口頭告知資訊，則可約定在揭露後多久的期限內以書面告知機密資訊的指定。此種

定義方法的好處是不會產生認知上的誤差，缺點則是，若提供大量資料，一一標示或書面指定，顯然很費工。而且若以口頭揭露，在揭露時點到書面說明機密性的時點之間，若已對外揭露，可能引發爭執。有時候，收訊方會要求揭露方，必須事先以書面提示將揭露哪些機密資訊，再真的揭露機密資訊。因為接收機密資訊而負擔保密義務，會使得收訊方行動受限。可能發生以不重要的機密資訊綁住收訊方行動的情況。（註3）此種定義的寫法如下：

例句二：（註4）

"Confidential Information" means

(a) any information about Discloser or its business

    (i) stamped "confidential" or

    (ii) within x days after disclosure identified in writing as confidential to the Recipient or any of its Representatives by or on behalf of the Discloser at the time of or promptly following the information's written or oral disclosure; and

(b) all notes, analyses, compilations, studies, summaries, and other materials, however documented, containing or based, in whole or in part, on any information described in subsection (a)（collectively, the "Derivative Materials"）

"Representatives" means, with respect to any persons, its

---

3   Marcel Fontaine & Fillip De Ly, *Drafting International Contracts: An Analysis of Contract Clauses* 246,247 (Martinus Nijhoff 2009).

4   Michael A. Woronoff, *Confidentiality* in *Negotiating And Drafting Contract Boilerplate* 406 (Tina L. Stark et al. ed., ALM Publishing 2003).

directors, officers, employees, agents, consultants, advisors, or other representatives.

3. 混合式

由於前兩種方法各有優缺點，因此也可加以混合併用。寫法如下

例句三：（註5）

Confidential Information.

As used in this Agreement, "Confidential Information" shall mean all tangible and intangible non-public information in any form（including written information, oral statements, visual observations of Discloser's operations at its business premises, and electronically stored data）disclosed by the Discloser (i) that (a) deriving independent economic value, actual or potential, from not being generally known to, and not being readily ascertainable by proper means by, other persons who can obtain economic value from its disclosure or use, and (b) is the subject of efforts that are reasonable under the circumstances to maintain its secrecy", or (ii) all other information (a) identified by the Discloser as confidential or proprietary, or (b) deemed confidential, protected, a trade secret, or proprietary under applicable law.

5　International Trade Centre UNCTAD/WTO, *Confidentiality Agreement*, http://www.jurisint.org/doc/orig/con/en/2001/2001jiconen73/2001jiconen73.pdf, accessed on 2011/11/21

4. 排除約定

　　某些資訊雖然符合以上的正面定義，但可能被排除於機密的範圍之外，例如已經合法公開，或者收訊方已合法掌握訊息，或嗣後由第三人處合法取得的訊息，因此需要加以排除。

例句四： （註6）

Despite the definition of "Confidential Information" set forth in subsection (a), "Confidential Information" excludes information that the Recipient may demonstrate

(i)　was or becomes generally publicly available other than as a result of a disclosure by Recipient or any of its Representatives in violation of this Agreement;

(ii)　is in the lawful possession of the Recipient or any of its Representatives prior to its disclosure by or on behalf of the Discloser or any of its Representatives; or

(iii)　was or becomes available to the Recipient or any of its Representatives on a nonconfidential basis from a third party that is not bound by a similar duty of confidentiality（contractual, legal, fiduciary, or other）.

　　有時候接受資訊的一方會要求加入豁免「殘存記憶」（residual memory）的約定，也就是說，接受訊息一方的員工還是可以將記憶中的資訊拼裝起來進行利用。接受資訊的一方考量的是，員工腦子裡可能

6　改編自 Michael A. Woronoff 前揭著第409頁。

記得許許多多斷簡殘編的資訊，若在研發過程中，只因為被記憶中的片斷資訊污染了，而造成整個研發成果無法使用，那是很可怕的。當然，對提供資訊的一方來說，這種約款聽起來也是很恐怖，只要接受資訊方派幾個一目十行、過目不忘的奇人異士來打探軍情，機密資訊就全都破功了。

　　因此這種條款也是猛烈交火的議題。在雙方交火之後，若非全有或全無的情況，可能有一些妥協的方法，例如這類豁免約款，只適用在業內人士通常具備的知識技能，不得借助任何機器設備或其他外力。若要加入殘存記憶，可在範例四的最末加上第（iv）款。

例句五：

　　(iv) is Residual Information retained in non-tangible form （i.e., in a person's memory）without aid of any equipment by Representatives of Recipient who rightfully had access to such information and who do not make an effort to purposely retain such information in memory in order to avoid the field of use restrictions otherwise applicable.

第三節　許可使用或揭露事由

　　提供機密資訊之目的就是要運用，但運用有其範圍限制，需要加以約定。而接受資訊方在盡契約義務，享契約權利時，無可避免地，也需要進行內部與外部的資訊傳遞，此時即需要定出相關的規矩。

例句六：^{（註7）}

Permitted Use. The Recipient may use Confidential Information only to perform under or receive the benefits of this Agreement.

Permitted Discloses. The Recipient may disclose Confidential Information to only those of its Representatives who

(a)　require the Confidential Information for the Permitted Use but to the extent practicable, only the part that is required;

(b)　are informed in writing by the Recipient of the confidential nature of the Confidential Material; and

(c)　agree in writing to be bound by the obligations of this Article.

第四節　對機密資訊負擔的義務

定義何為機密資訊之後，即應說明對於機密資訊應負擔哪些義務。雖然我們泛稱保密義務，但具體內容必須記明，絕對不能只寫「應該保密」，因為那太過籠統，實際操作時容易引發糾紛。也不必約定「鎖在上鎖的櫃子裡，要有何種加密措施」，因為那太過繁瑣，容易掛一漏萬。保密義務的重點通常包括：(1)非經許可，不得做指定用途之外的利用。(2)非經許可，不得揭露給第三人。(3)政府機關強制揭露時的因應措施。(4)銷毀或返還資料。

7　Michael A. Woronoff 前揭著第 412，413，414，422 頁。

1. 不得作指定用途之外的利用

　　有時除了約定不得作指定用途外之利用外，還會特別例示說明不能做的事，例如還原工程。還原工程一般是合法的，所以在契約中特別約定不能做還原工程，是否有效，還是需要依照個別的準據法個案判斷。

例句七：

　　Except for Permitted Use, the Recipient shall not use Confidential Information for any purpose. To the fullest extent permitted under applicable law, the Recipient shall not modify, disassemble, decompile, adapt, alter, translate, reverse engineer, or create derivative works based on the Confidential Information, or any materials associated or included with, or embedded into, the Products.

2. 不得揭露給第三人

　　我們直接以例句解說：

例句八：

　　Except for Permitted Disclosure, the Recipient shall not disclose Confidential Information to any third party.

3. 政府機關強制揭露時的因應措施

　　有時候政府部門（例如法院、檢調單位）會介入，而命令收訊方提供機密資訊，因此有必要約定收訊方收到此種命令時應如何應對，包括通知、協商，提出異議的程序。

例句九： (註8)

Compelled Disclosures.

(a) *Notification, Consultation, and Protective Orders*. If the Recipient or any of its Representatives is requested, in any case by a court or governmental body, to make any disclosure of Confidential Information, the Recipient shall

(i) promptly, but in any event no later than __ days after the Recipient becomes aware that it is required to make such disclosure, notify the Discloser in writing;

(ii) consult with and assist the Discloser in obtaining an injunction or other appropriate remedy to prevent such disclosure; and

(iii) use its reasonable efforts to obtain at the Discloser's expense a protective order or other reliable assurance that confidential treatment will be accorded to any Confidential Information that is disclosed.

(b) *Right to Disclose*. Subject to the provisions of Section (a), the Recipient or its Representative may furnish that portion

---

8　改編 Michael A. Woronoff 前揭著第 423 頁例句。

and only that portion of the Confidential Information that the Recipient or its Representative is legally compelled or otherwise required to disclose.

4. 銷毀或返還資料

在雙方合作關係終止時，或者在特定情況下經由揭露方要求時，收訊方必須返還機密資料或予以銷毀。

例句十：（註9）

Return and Destruction of Confidential Information. Upon the termination of this Agreement or written request of the Discloser, the Recipient shall, and shall cause its Representatives to promptly, but in any event no later than ___ days after termination or the Recipient's receipt of the Discloser's written request,

(i) return to the Discloser all Confidential Information furnished to the Recipient or any of its Representatives; and

(ii) destroy all Derivative Material and upon such destruction, the Recipient shall certify in writing to the Discloser that it has done so.

9　改編Michael A. Woronoff前揭著第423頁例句。

第五節　對機密資訊負擔保密義務的期限與方法標準

　　如果是單獨簽訂保密契約，可能獨立設定保密期限，但如果是在各種契約中加上保密條款，則保密期限可能是契約到期後再加一段時間。而保密標準可能是商業上合理、或與處理自己事務相近注意，產業相關實務，或是這些標準的混合。

例句十一：（註10）

Obligation to Maintain Confidentiality. During the term of this Agreement and within __years after this Agreement expires or is terminated, the Recipient shall take all commercially reasonable measures necessary to keep the confidentiality of Confidential Information including, without limitation, all measures it takes to protect its confidential information of a similar nature. Without limiting the effect of the preceding sentence, the Recipient shall take commercially reasonable actions, legal or otherwise, necessary to cause its Representatives to comply with the provisions of this Agreement and to prevent any disclosure of the Confidential Information by any of them.

Unauthorized Use. The Recipient shall give prompt written notice to the Discloser of any unauthorized use or disclosure of the Confidential Information and shall assist the Discloser in remedying each unauthorized use or disclosure. Giving

---

10　改編 Michael A. Woronoff 前揭著第 412 頁例句。

assistance does not waive any breach of this Article by the Recipient, nor does acceptance of the assistance constitute a waiver of any breach of this Article.

第六節　違約救濟

　　當事人可以約定若收訊方違反保密約定，揭露方得終止或解除契約、請求損害賠償、違約金、轉嫁對第三人的責任，或者向法院或其他政府機關聲請相關處分、禁制措施，這些救濟方法可能歸併在終止、解除條款，賠償條款、責任轉嫁條款與其他救濟條款，也可能獨立地列在保密條款中，若要列在保密條款，寫法如下。在聲請假處分的程序中，需特別留意是否同意聲請人免提擔保金，且免提出證據，而填補損失或轉嫁責任的範圍除了揭露方之外，是否還包括揭露方的董事、高階主管與關係企業、甚至客戶，這些可能是雙方爭執的重點。

例句十二：（註11）

Injunctive Relief. The Recipient acknowledges and agrees that because

(a)　an award of money damages is inadequate for any breach of this Agreement by the Recipient or any of its Representatives, and

(b)　any breach causes the Company irreparable harm, in the event of any breach or threatened breach of this Article by

---

11　改編 Michael A. Woronoff 前揭著第424頁例句。

the Recipient or any of its Representatives, the Discloser is entitled to equitable relief, including injunctive relief and specific performance, without the posting of a bond or other security and without proof of actual damages.

Indemnity. The Recipient shall indemnify and defend the Discloser and its Affiliates against all damages, losses, costs, liabilities and expenses including reasonable legal fees and the cost of enforcing this indemnity, arising out of or relating to any unauthorized use or threatened use or disclosure or threatened disclosure by the Recipient or any of its Representatives of the Confidential Information or any other violation of this Article.

第九章

競業禁止條款與
獨家交易條款

　　有時候契約只約定保密條款是不夠的，因為要證明對方洩密，或對機密資訊做許可用途以外的利用，並不容易。所以有時會更進一步地禁止對方從事競爭的商業活動，例如在僱傭契約中禁止員工在任職期間內，或離職後一段時間內從事競業活動，在併購契約中禁止與被併標的之股東從事競業活動，在經銷代理或加盟連鎖契約中限制經銷商，或代理商經銷或代理其他品牌的競爭產品。

　　但是競業禁止的條款可能受到多種法律的限制，例如基於保護勞工權益的法令，例如基於保護消費者與競爭秩序的競爭法令。所以禁止競業要有合理的目的，範圍也必須合理。有時候法令雖未禁止限制競業，但如果約定了限制競爭條款，可能有補償義務，例如對受僱人的補償，或者歐盟及會員國規定供應商對於代理商的補償義務[註1]。

　　另外還有一種獨家交易條款與競業限制效果相近，但限制的對象行為不同。競業限制是不能做跟對方類似的營業或銷售類似的商品，但獨家交易則是只能與特定交易對象從事某種買賣交易、不能與其他任何人從事相似的交易。比方說，在經銷契約中，競業限制針對的行為是經銷商銷售其他廠牌產品，所以經銷商可以買其他廠牌的產品自己使用。但獨家交易的重點，則是經銷商只能向特定供應商買貨，不能向別人買貨，所以經銷商可以自己製造類似產品來賣。獨家交易通常適用於供應契約、經銷契約。在供應契約中，獨家交易限制的對象可能是買方，也可能是賣方。另外在併購意向書或財務規劃契約，還有期間較短的類似限制條款，則常稱為 standstill obligations。以下我們逐一說明。

1　例如歐盟歐盟 COUNCIL DIRECTIVE of 18 December 1986 on the coordination of the laws of the Member States relating to self-employed commercial agents（86/653/EEC)(Agency Directive）第17條要求各會員國立法保護代理商在代理關係終止或到期後獲得補償，而德國商法§89b第1項與瑞士債法418u都有類似的規範。歐盟會員國有些也將補償代理商的規範類推適用到經銷商，所以在簽訂相關契約時，應特別留意此類補償規範。

第一節　競業禁止條款

1. 被禁止或限制之人

　　競業禁止條款適用的交易類型多元多樣，在不同交易中，被禁止或限制之人也不同。在併購契約或股份買賣契約中，受限制的對象可能是被收購股份的股東，或合併後消滅公司的股東．董監事等等，在僱傭契約中則是受僱人，在經銷契約中則是經銷商，在代理契約中則是代理商。

2. 限制的正當性

　　有時在契約中也會加入一段文字說明限制的正當性，這段文字有助於雙方當事人先想想這樣的限制是否合理，若確認合理，將判斷合理的基礎前提事實明文列出，可以避免未來一方反口爭執基礎事實，也可以避免在爭訟中，遺忘了原先約定的合理性基礎。

3. 競業禁止的範圍

（1）時間：

　　以受僱人來說，可分為在職期間，以及離職後特定期間的競業禁止。在職期間禁止競業是天經地義的，較無問題。但離職後，綁多久不會失去合理性，就需要斟酌了。以臺灣的勞動基準法為例，即限制最長不能超過離職後二年。

（2）受限制的競業行為：

以受僱人的競業限制來說，受限制的行為包括自己經營競爭事業、或參與第三人的競爭事業，無論是有償或無償，無論是受僱、承攬或委任、無論擔任員工、顧問、董事、監察任或任何頭銜的職務。除此之外也要留意投資股份也可能產生利益衝突的情況，當然買賣上市股票不應有過嚴苛的限制，避免顯失公平合理。有的也會明文限制離職員工不能勸誘同事離職。

4. 通知義務

有的僱主在競業禁止契約，或競業禁止條款中，並非全面禁止，而是要求需取得僱主事先書面同意，而且也要求員工離職後在限制期間內若要接受另一份工作，也要提供新僱主，該員工已與老東家簽定的競業禁止或限制約款。

5. 違反效果

以受僱人來說，違反競業禁止義務，若是在職期間內，可能構成終止僱契約的事由。而無論是在職期間內違反義務或離職後違反義務，均可能需要賠償違約金[註2]，甚至採取類似公司法對董事違反不競業義務時的利益歸入設計，就是員工違反競業禁止義務而獲得任何利益，僱主可基於不當得利而要求員工將獲益移轉給僱主。也有的會約定同意可聲請禁制令。

2　需留意在，若適用英美法，不容許懲罰性違約金，只能要求預定損害數額的違約金（liquidated damage）。

6. 補償

如前所述，對員工、對代理商，法律可能規定設定競業禁止或限制條款的一方必須補償受限制的對象。

以下我們提供競業競爭條款實例：

例句一：（註3）

1. ACKNOWLEDGEMENTS

(a) The Employee acknowledges that the Company (as defined below) has expended and shall continue to expend substantial amounts of time, money and effort to develop business strategies, employee and customer relationships and goodwill, and to build an effective organization. The Employee acknowledges that the Company has a legitimate business interest in protecting such efforts. The Employee also acknowledges and agrees that in the performance of the duties and responsibilities of employment, the Employee has and will become familiar with the Company's confidential information, including trade secrets, and that the Employee's services are of special, unique and extraordinary value to the Company. Further, the Employee acknowledges that the Company would be seriously and irreparably damaged by the disclosure of trade secrets, confidential information and/or the loss or deterioration

---

3　美國證管會上市公司申報NON-COMPETITION AND NON-SOLICITATION AGREEMENT，https://www.sec.gov/Archives/edgar/data/1075531/000107553113000016/exhibit991non-competeagree.htm, 最後點閱日：August 2，2023。

of its business strategies, employee and customer relationships and goodwill. The Employee further understands and agrees that the foregoing makes it necessary for the protection of the business and the Company that the Employee not compete with the Company during his or her employment and for a reasonable period thereafter, as further provided in the following Paragraphs.

For purposes of this Agreement:

(i) Affiliate" means, with respect to any specified person or entity, any other person or entity that directly or indirectly, through one or more intermediaries, Controls, is Controlled by, or is under common Control with, such specified person or entity; and

(ii) Company" means priceline.com Incorporated as well as all direct and indirect subsidiaries and Affiliates of priceline.com Incorporated;

(iii) Control" (including, with correlative meanings, the terms "Controlled by" and "under common Control with") , as used with respect to any person or entity, means the direct or indirect possession of the power to direct or cause the direction of the management or policies of such person or entity, whether through the ownership of voting securities, by contract or otherwise.

(b) The Employee acknowledges (i) that the business of the Company is global in scope and without geographical limitation and (ii) notwithstanding the jurisdiction of formation or principal office of the Company or any of its respective executives or employees（including, without limitation, the Employee）, it is expected that the Company will have business activities and valuable business relationships within its industry throughout the world. In addition, the Employee agrees and acknowledges that the potential harm to the Company of the non-enforcement of the provisions of this Agreement outweighs any potential harm to the Employee of its enforcement by injunction or otherwise.

(c) The Employee acknowledges that (i) he or she has carefully read this Agreement, fully understands the restraints imposed upon the Employee by this Agreement, and voluntarily agrees to its terms and conditions; (ii) he or she was not coerced to sign this Agreement and was not under duress at the time he or she signed this Agreement; (iii) by signing this Agreement, he or she will not violate the terms of any other agreement previously entered by the Employee; and (iv) prior to signing this Agreement, he or she had adequate time to consider entering into this Agreement, including, without limitation, the opportunity to discuss the terms and conditions of this Agreement, as well as its legal consequences with an attorney of his or her choice. The Employee expressly acknow-ledges and agrees that, especially given the nature and scope of the Company's business, each and every restraint imposed by this Agreement is reasonable with respect to subject matter, time period and geographical area.

2. NONCOMPETITION AND NON-SOLICITATION

(a) The Employee will not, while an employee of the Company, and for a period of one year following the termination of his or her employment（the "Restriction Period" as further defined below）, directly or indirectly, without the prior written consent of the Company:

(i) (A) engage in any of the same or substantially similar activities, duties, or responsibilities in the line of business or relating to the line of business that the Employee had responsibility for or knowledge of while an employee of the Company, for any other company that competes with such line of business of the Company, including any of the following companies or their successors: (I) Expedia, Hotels.com, Hotwire and Venere; (II) Sabre Group and Travelocity; (III) Lastminute.com plc; (IV) Travelport, including, without limitation, Orbitz, CheapTickets, Lodging.com, the Neat Group and Galileo; (V) the following on-line travel aggregators: Mobissimo, Inc. (owner and operator of the website Mobissimo.com), Cheapflights Limited (owner and operator of the website Cheapflights.com), Farechase, Kayak.com, Trivago, Tripadvisor, or any substantially similar on-line travel search business; (VI) C-trip; (VII) Wotif; (VIII) HRS; and (IX) Roomkey; and (X) the on-line travel search businesses of Yahoo!, MSN, AOL or Google;

(B) solicit or attempt to solicit any customer or client, or actively sought prospective customer or client, of the Company with respect to the businesses actively operated by the Company, to purchase any travel related goods or services of the type sold by the Company from anyone other than the Company（it being intended that businesses owned but not operated by the Company, such as, as of the date hereof, the Company's mortgage business, are not to be covered by this clause (B)）；or

(C)assist any person or entity in any way to do, or attempt to do, anything prohibited by (A) or (B) above; or

(ii) (A)solicit, recruit or hire to work for the Employee or any organization with which the Employee is connected, any employees of the Company or any persons who, within one (1) year of such solicitation, recruitment or hire, have worked for the Company;

(B) solicit or encourage any employee of the Company to leave the services of the Company; and

(C) intentionally interfere with the relationship of the Company with any person who is employed by or otherwise engaged to perform services for the Company; provided, that neither (I) the general advertisement for employees or the general solicitation of employees by a recruiter, nor (II) the Employee's being named as an employment reference for a current or former employee of the Company and responding to ordinary course inquiries made of the Employee by prospective employers of such employee in connection with such reference, shall be deemed a violation of this clause (ii).

The "Restriction Period" means the one-year period following the cessation of the Employee's employment with the Company for any reason. The Restriction Period shall be tolled during（and shall be deemed automatically extended by）any period in which the Employee is in violation of the provisions of this Section 2.

(b) Notwithstanding anything to the contrary contained in this Agreement, the foregoing covenant will not be deemed breached as a result of the Employee's passive ownership of less than an aggregate of 5% of any class of securities of any entity listed in Section 2(a)(i)(A);provided, however, that such stock is listed on a national securities exchange or is quoted on the National Market System of NASDAQ.

(c) If any provision or clause of this Agreement, or portion thereof , is held by any court or other tribunal of competent jurisdiction to be illegal, invalid, unreasonable, or otherwise unenforceable against the Employee, the remainder of such provision shall not be thereby affected and will be deemed to be modified to the minimum extent necessary to remain in force and effect for the longest period and largest geographic area that would not constitute such an unreasonable or unenfor-ceable restriction. It is the express intention of the parties that, if any court or other tribunal of competent jurisdiction construes any provision or clause of this Agreement, or portion thereof, is held by any court or other tribunal of competent jurisdiction to be illegal, invalid, unreasonable, or otherwise unenforceable against the Employee because of the duration of such provision, the scope of the subject matter, or the geographic area covered thereby, such court or tribunal shall reduce the duration, scope, or area of such provision, and, in its reduced form, such provision shall then be enforceable and be enforced. Moreover, notwithstanding the fact that any provision of this Section 2 is determined not to be enforceable in equity, the Company will nevertheless be entitled to recover monetary damages as a result of the Employee's breach of such provision.

3. NOTIFICATION OF SUSEQUENT EMPLOYER

The Employee hereby agrees that prior to accepting employment with any other person or entity during any period during which the Employee remains subject to any of the covenants set forth in Section 2, the Employee shall provide such prospective employer with written notice of such provisions of this Agreement, with a copy of such notice delivered simultaneously to the Company.

4. REMEDIES AND INJUCTIVE RELIEF

The Employee acknowledges that a violation by the Employee of any of the covenants contained in Section 2 would cause irreparable damage to the Company in an amount that would be material but not readily ascertainable, and that any remedy at law（including the payment of damages）would be inadequate. Accordingly, the Employee agrees that, notwith-standing any provision of this Agreement to the contrary, the Company shall be entitled（without the necessity of showing economic loss or other actual damage）to injunctive relief （including temporary restraining orders, preliminary injunctions and/or permanent injunctions）in any court of competent jurisdiction for any actual or threatened breach of any of the covenants set forth in Section 2 in addition to any other legal or equitable remedies it may have. The preceding sentence shall not be construed as a waiver of the rights that the Company may have for damages under this Agreement or otherwise, and all of the Company's rights shall be unrestricted.

5. CONSIDERATION

In consideration for the Employee's agreement to comply with this Agreement and covenants herein, the adequacy of which is hereby acknowledged by the Employee, the Company has granted the Employee an equity award under the priceline. com Incorporated 1999 Omnibus Plan, as amended, and an award agreement thereunder（the "Equity Award Agreement"）as well as continued employment with the Company. A material breach by the Employee of any of his or her obligations under this Agreement shall be considered grounds for a "Termination for Cause" pursuant to the terms of the Equity Award Agreement.

第二節　獨家交易條款

　　獨家交易條款有許多類型，可能適用在智慧財產權的獨家授權、可能適用在商品交易上的獨家安排。智慧財產權上的獨家授權，通常合法性較無疑慮。而商品交易上的獨家安排可能限制買方只能向賣方買特定

標的，不能向第三人買，也可能限制賣方只能向買方供應特定標的，即不能向第三人供貨。獨家交易對整體競爭情勢可能有利有弊，上下游專業分工、垂直整合，固然可以提升效率、改善品質，但也可能排除競爭者，而產生限制競爭的效果。各法域對於獨家交易條款有不同的規範要求、執法立場與審查方法。本書難以提供一體適用的遵法建議，讀者必須在個案中審酌法令規範。但建議必須先理解獨家交易條款，可能的法律風險，在條款中也預先設計風險的因應方式。通常限制買方不得向其他廠商購買的違法風險，高於限制賣方只能賣給特定買方。無論獨買或獨賣，通常也都會綁定最低購買量或最低銷售量，否則承諾獨買或獨賣，卻未達到預期的量，就會有重大損失。以下我們提供實例給讀者參考。

例句二：（註4）

2. SALE AND PURCHASE OF MATERIALS

2.1Supply of Material.

(a) SiTech hereby agrees to manufacture for, and deliver exclusively to Mentor, and Mentor agrees to purchase from SiTech, such quantities of the Materials to meet Mentor requirements based upon such written purchase orders and forecasts provided Pursuant to Section 2.3 hereof. In the event that Mentor's requirements differ significantly (by more than 10%) from the forecasts, Mentor will promptly notify SiTech of the fact and of the amount of such variance, and SiTech shall

---

4　美國證管會上市公司申報Exclusive Supply AGREEMENT，https://www.sec.gov/Archives/edgar/data/64892/000006489204000014/ex-sitechag.htm 最後點閱日：August 4，2023。

use its best efforts to accommodate any variance upon receiving notice thereof from Mentor. The parties hereto acknowledge and agree that SiTech may manufacture or sell other products to any third party upon the prior written consent of Mentor, which consent may be withheld by Mentor at Mentor's sole and absolute discretion.

(b)Mentor shall be provided with a list of raw materials and finished goods inventory on a quarterly basis. Such list shall reflect the status of SiTech's inventory at the end of each calendar quarter sufficient to assure that Mentor shall have a three (3) month supply of the materials to be purchased by Mentor pursuant to the terms of this Agreement (the "Minimum Inventory Level"). Mentor shall have the right, but not the obligation, to conduct an annual audit of SiTech, at Mentor's expense, to satisfy itself that such Minimum Inventory Level is being met. In the event that Mentor's audit or any SiTech Quarterly report shall reveal that SiTech has not maintained the Minimum Inventory Level for each necessary raw material, Mentor may, at its sole option, purchase a three (3) month supply of any such raw material and store such raw material at SiTech's facility. SiTech shall reimburse Mentor for Mentor's actual cost for such raw materials upon SiTech's use of the raw material purchased by Mentor.

另外也補充說明獨家經銷條款（sole distributorship）與專屬經銷條款（exclusive distributorship）。這兩種條款的差異在於獨家經銷，是供應商承諾在指定區域內不授權第二家經銷商，但未承諾供應商不自

己直銷。而專屬經銷則是供應商承諾在指定區域內，既不授權第二家經銷商，供應商也不會自己直銷。獨家（sole）與專屬（exclusive），不論用英文或中文，都很難顧名思義理解成上述任一種意思，甚至不同的使用者混用兩個詞語，所以我們建議在契約中要講清楚，到底限制內容是什麼，避免雙方認知有落差而產生糾紛。獨家經銷與專屬經銷既有限制競業的意思，也有獨家交易的意思。

雖然口語上常常說授與經銷權，但事實上法律並未規定經銷權這樣的權利，專利可授權、商標可授權，但法律上沒有經銷權這種東西。經銷權究竟是什麼？沒有經銷權就真的不能賣嗎？到 A 國跑單幫買貨拿到 B 國賣，沒有供應商授權，可以嗎？法律上沒有說不可以。尤其如果涉及智慧財產權，還有所謂的權利耗盡原則，買家買了有正版智財權的商品，就可以自己轉賣，除非正版智財權產品區分成出租版、家用版、公播版，而限制買方行使享受權利的方法，則買方只能將自己有的權利再售出，不能出賣自己沒有的權利。

因此，銷售權實際上並無法完全壟斷，就算授與 A 公司銷售權，B 公司也可能從 A 公司或其他管道取得貨源而轉售。甚至許多法域的競爭法針對廠商干擾平行輸入的行為設有罰則。那為什麼我們常聽到授與經銷權呢？實際上，經銷權背後的支持基礎，常常在於使用商標或其他智慧財產的權利，或者供應商以提供原廠售後服務的方式來支持經銷商的銷售權。所以經銷契約中也必須思考經銷商的經銷權獲得哪一種支持。

探討經銷是不是權利的目的，在於辨明我們在契約中對獨家經銷與專屬經銷的內容該如何敘述、轉譯成法律文字，寫成 A 公司授予 B 公司經銷權利是有問題的，寫成 A 公司委任 B 公司作為獨家經銷商也是常見的做法，但也有問題。如果是非獨家也非專屬的經銷，這種經銷契約事實上與一般的買賣契約差異不大，至多可能經銷關係多了一些共同行銷

推廣費用的分攤，以及技術協助或客訴處理的約定。契約的關鍵核心在於設定權利與義務。權利的重點就是若無授權就不能做，但以上已經說明，經銷並不是這樣，可以平行輸入，可以賣水貨。義務的重點在於義務人必須完成某種事情。所以在經銷契約中，供應商委任獨家經銷商銷售，可能課予經銷商每年要賣到某一種量，但未必在所有情況都能談成買不足就要補貨，常常是買不足量，供應商可以終止或不續約，或改成非獨家、非專屬。所以這樣的義務拘束仍然較弱，而且經銷商賺錢不是因為供應商直接發錢給經銷商，如果是直接發錢，那是代理關係，而非經銷。經銷商是透過轉賣的價差而賺錢。所以若將經銷關係解釋為委任關係，那樣的義務拘束較弱。

　　總之，非獨家、非專屬的經銷，我們可以理解為就是買賣關係。如果是獨家或專屬的經銷，其重點在於限制供應商，而不在於限制經銷商。當然供應商受限制是以經銷商也受某些限制（例如最低購買量）為前提條件。而供應商受到的限制，就是**在指定區域內，供應商不能賣給其他經銷商（獨家經銷）、或供應商除了不能賣給其他經銷商之外，也不能賣給最終用戶（專屬經銷）**。所以在獨家經銷或專屬經銷中，必須明明白白地寫出這樣的限制，否則法官未必會替當事人從單純的獨家經銷，或專屬經銷這幾個字推演出限制供應商的內容。在許多大陸法系的國家，法律甚至也未對經銷關係作出明文定義，就像在臺灣，經銷契約可能被解釋為委任關係，或解釋成無名契約，委任關係對委任人並沒有太大限制，委任人原則上可以　時再委任別人，委任人原則上也可以自己辦理委任事務。所以在獨家經銷或專屬經銷契約中，若未寫明關鍵限制，發生糾紛時，法官做出的裁判可能不如經銷商之預期。

　　以下我們提供範例給讀者參考：

例句三：

3. Exclusive Distributorship.

The Supplier may neither sell Products to any buyer located in the Territory no matter for purpose of transfer with or without charge, or for purpose of own consumption, nor sell Products to any buyer located outside the Territory if the Supplier knows such buyer will transfer Products to any one located in the Territory with or without charge.

例句四：

3. Sole Distributorship.

The Supplier may not sell Products to any other distributor located in the Territory, nor sell Products to any other distributor located outside the Territory if the Supplier knows such distributor will resell Products to any one located in the Territory.

第三節　Stand Still 條款

Stand Still 字面上的意思就是站住別動。Stand Still 條款常用在財務顧問提供募資規劃，或者磋商新股發行或老股買賣，或甚至高額不動產買賣的情況，這種條款的邏輯在於打算交易的一方，會因為這樣的交易磋商而負出高昂的成本，例如財務顧問幫當事人規劃募資方案，投入許多人力，如果客戶自己接觸到其他管道，讓財務顧問白忙一場，豈非虧

大了。又如新股發行或老股買賣、或高額不動產買賣，可能的買方接下來可能要委託律師進行複雜的盡職調查，這樣的工作都是要花很多錢的，所以如果不把賣方綁起來，讓賣方還可以招徠別的買家，做盡職調查的買家就會白花錢了。

Stand Still 的重點就在於綁的對象、綁的期間（包括可能的伸縮）、限制禁止的行為、違約的賠償。以下我們提供實際的例句給讀者參考。

例句五：（註5）

1. Standstill Provision. Securityholder agrees that from the date of this Agreement through the earlier of (i) the second anniversary of the expiration of the Lock-Up Period (as defined below), and (ii) the date that the Securityholder (together with its Affiliates (as defined below)) no longer beneficially owns Common Stock (including shares underlying options or warrants) representing, on an as converted basis, in the aggregate, at least 10% of the Company's outstanding Common Stock (making equitable adjustments for any conversions, reclassifications, reorganizations, stock dividends, stock splits, reverse splits and similar events which occur with respect to the Common Stock), neither the Securityholder nor its Affiliates will, directly or indirectly, without the prior written consent of a majority of the Board of Directors of the Company (excluding any nominees or designees of the Securityholder on the Board

---

5 美國證管會上市公司申報 Lock-Up and Standstill Agreement, https://www.sec.gov/Archives/edgar/data/879407/000119312507090245/dex103.htm，最後點閱日：August 5，2023。

of Directors), in the directors' sole and absolute discretion, acquire, agree to acquire, make any proposal to acquire, or in any way participate in a "group" (within the meaning of Section 13(d)(3) of the Securities Exchange Act of 1934, as amended, (the "Exchange Act")) to do any of the foregoing, equity securities (including convertible debt instruments and preferred stock or any shares of capital stock issuable upon the conversion or exercise thereof) of the Company representing more than 20% of the voting power of all voting securities of the Company on a fully diluted basis. "Affiliate" means, with respect to any specified person, a person that directly, or indirectly through one or more intermediaries, controls, or is controlled by, or is under common control with, the specified person, where "control" (including the terms "controlling," "controlled by" and "under common control with") means the possession, directly or indirectly, of the power to direct or cause the direction of the management and policies of such person, whether through the ownership of voting securities, by contract, or otherwise.

2. General Lock-Up of Securities. Subject to the provisions of Section 1 hereof and the last sentence of this Section 2, and subject in all events to the consummation of the Stock Purchase Transaction, during the period commencing on the date of this Agreement and ending on the One Hundred and Eighty-First (181st) day following the date of this Agreement (such period is referred to herein as the "Lock-Up Period"), the Securityholder shall not (a) sell, transfer, assign, offer, pledge, contract to sell, transfer or assign, sell any option or contract to purchase,

purchase any option or contract to sell, transfer or assign, grant any option, right or warrant to purchase, or otherwise transfer, assign or dispose of, directly or indirectly, any of (1) the Shares or (2) any other securities of the Company now held or hereafter acquired by the Securityholder, including, but not limited to, securities convertible into or exercisable or exchangeable for Common Stock (all such securities, the "Locked-Up Securities"), (b) enter into any swap or other arrangement that transfers or assigns to another person or entity, in whole or in part, any of the economic benefits, obligations or other consequences of any nature of ownership of the Locked-Up Securities, whether any such transaction is to be settled by delivery of the Locked-Up Securities in cash or otherwise, or (c) engage in any short selling of the Common Stock or securities convertible into or exercisable or exchangeable for Common Stock. Each of the transactions referred to in the foregoing clauses (a), (b) and (c) is referred to herein individually as a "Sale" and collectively as "Sales." Notwithstanding the restrictions set forth in this Section 1, during the Lock-Up Period the Securityholder shall be permitted (i) to transfer Locked-Up Securities to transferees for estate planning purposes and (ii) to transfer or sell Locked-Up Securities in privately negotiated transactions that are exempt from the registration requirements of the Securities Act of 1933, as amended, and from the registration and qualification requirements of all applicable state securities laws, so long as under either clause (i) or (ii) any such transferee or purchaser, as the case may be, signs a written instrument satisfactory to the Company in its sole discretion evidencing such transferee's or purchaser's agreement to be bound by the provisions of this Agreement with respect to the Locked-Up Securities so transferred.

3. Trading Plans. Notwithstanding anything herein to the contrary, nothing herein shall prevent the Securityholder from establishing a 10b5-1 trading plan that complies with Rule 10b5-1 under the Exchange Act or from amending an existing 10b5-1 trading plan so long as (i) there are no sales or dispositions of securities of the Company under such plans during the Lock-Up Period, as the same may be extended hereby and (ii) no party, including the undersigned, shall be required to, nor shall it voluntarily, file a report under Section 16(a) of the Exchange Act in connection with the adoption or amendment of such trading plan.

4. Extension of Lock-Up Period. Notwithstanding the foregoing, if (1) during the last 17 days of the Lock-Up Period, the Company issues an earnings release or material news or a material event relating to the Company occurs; or (2) prior to the expiration of the Lock-Up Period, the Company announces that it will release earnings results during the 16-day period beginning on the last day of the Lock-Up Period, the restrictions imposed by this Agreement shall continue to apply until the expiration of the 18-day period beginning on the issuance of the earnings release or the occurrence of the material news or material event.

5. Equitable Remedies. The Securityholder acknowledges

and agrees that the Company's remedy at law for a breach or threatened breach of any of the provisions of this Agreement would be inadequate and, in recognition of this fact, in the event of a breach or threatened breach by the Securityholder of any of the provisions of this Agreement it is agreed that, in addition to its remedy at law, the Company shall be entitled, without posting any bond, to equitable relief in the form of specific performance, temporary restraining order, temporary or permanent injunction or any other equitable remedy which may then be available. Nothing herein contained shall be construed as prohibiting the Company from pursuing any other remedies available to it for such breach or threatened breach, including, without limitation, issuing stop transfer instructions to the Company's transfer agent in connection with any purported transfer of Locked-Up Securities by the Securityholder in violation of the provisions of this Agreement.

彌償條款與智慧財產權條款

第一節　彌償條款（Indemnity）

英文契約中常常可見到 Indemnity 條款，這個字在中文裡不易翻譯，意思就是預先承諾補償他人某種預期可能發生因為的損失，有風險承擔的意思。本書採用香港通常的譯法稱之為彌償條款。

個別契約中彌償條款的影響可能非常複雜，因為牽涉到訴訟制度、仲裁制度、甚至也牽涉到保險，基於彌償條款而獲得補償，可能會影響向保險公司求償的權利；或者保險公司可能要求轉讓，依據彌償條款得以主張的權利。

彌償條款基本上是一種給付義務，其結構簡言之就是，（1）彌償義務人，（2）應為受彌償權利人，（3）針對符合特定要件的第三人求償，（4）提供辯護及／或填補特定內容支出。我們以下逐一說明：

（1）彌償義務人：

彌償義務人必須是契約的一方當事人，張三跟李四簽約要王五買單是沒辦法執行的。

義務人可能是單數，也可能是複數。複數的時候，需特別處理這些義務人彼此間是各負各的責任，只能要 A 負擔 A 的部分，要 B 負擔 B 的部分（severally），或是共同負責，但必須一起被訴，請求全部數額，不能分開（jointly），或是依照權利人的選擇，可任意向債務人中的一人，或多人請求部分或全部金額（jointly and severally）。也要考慮是否需要求多個義務人指定一位代表代受送達等實際操作問題。

（2）受彌償權利人：

接受彌償的權利人可能只寫契約的當事人，也可能擴大包括契約當事人的股東、董事、員工、受任人、承攬人。

（3）彌償之原因事由：符合特定要件的第三人求償

應負彌償義務的狀況，最簡單的說，就是第三人求償。這裡必須特別說明，

彌償責任不應包括彌償權利人對義務人提出的主張，那是屬於損害賠償條款，而不是彌償條款。第三人為什麼能要求契約當事人賠償呢？通常是基於侵害智財權或基於產品責任。

權利人必然希望辯護義務與補償義務的範圍愈大愈好，但義務人希望盡可能限縮辯護義務與補償義務的範圍。如果彌償條款寫得不夠清楚，義務人可能負擔極重。契約當事人可以更進一步地詳細地，約定各種影響彌償義務範圍大小的要件。

有的約定只要第三人求償，義務人就必須負擔各種費用與支出，包括跟第三人談判、訴訟、仲裁的各種費用。有的約定必須第三人的主張，獲得終局裁判或仲裁確認，或者義務人盡早參與而同意的和解協議確認，義務人才負擔彌償義務。此外還可能有通知時間的限制，就像汽車保險，通常必須出險後24小時通知保險公司，契約中也可能限制第三人提出請求時，要在多久時間內通知權利人以利及早因應。也可能約定第三人求償發生原因的限制，例如第三人的求償是基於賣方供應商產品，侵害智慧財產權或賣方產品造成第三人人身傷亡或財產損害，或者第三人的求償，起因於或有關於彌償義務人違反契約的聲明擔保或給付義務。條款中也可能排除權利人故意或重大過失行為所造成的責任損

失，或者約定稅賦扣抵，或保險金的利益應由轉嫁額度中扣除（註1）

此外義務人也需考量，萬一權利人消極應訴結果輸了官司，或者甚至權利人跟求償的第三人串好，透過法律程序榨取義務人的賠償費用，該如何因應。因此，義務人可以在契約中要求由義務人指定權利人的辯護律師，而當然也可以約定彌償義務的上限金額。

（4）彌償義務之內涵：提供辯護，及／或填補特定內容支出

彌償是很特殊的條款，所以我們先區辨彌償與其他相似的概念，也就是擔保、連帶債務分攤、保證。

彌償與擔保（warranty）有其相似之處，但差別在於彌償強調一方承擔某種情況所造成的責任，而不是要求一方擔保絕對不會發生特定情況。有時候即使只寫擔保，還是可能被法院理解成是聲明且擔保。所以一方如果願意賠錢了事，但不想被告詐欺，或甚至因為法律的規定，而從聲明擔保推衍出解除或終止契約的權利，則彌償條款是較好的選擇。

Indemnity跟連帶債務分攤（contribution）也有差別（註2），contribution是多個債務人之間有一個責任分攤的比例。比如A、B各自開著自己的汽車，C走在路上，A車先把C撞飛；B車再把C輾過。假設C大難不死，狀告A、B兩人，A、B兩人之間有分攤責任的問題，A如果全賠了，可以向B要求分攤，這個是contribution，也就類似臺灣講的真正連帶債務。但如果情況是A公司把公司車交給B伙計開出去跑業務，B撞傷了C。後C起訴向A公司求償，A公司賠了錢後當然要回頭要求B全額償還。B有義務indemnify A公司，也就類似臺灣講的不

1　Morton A. Pierce & Michael C. Hefter, *Indemnities*, in *Negotiating And Drafting Contract Boilerplate* 257-309 (Tina L. Stark et al. ed., ALM Publishing 2003).
2　參閱前引 Pierce & Hefter 著作第 249 頁。

真正連帶債務。

在紐約州，假設消費者C向A經銷商買了B製造商生產的手機。回家以後手機電池燒掉，把C給燒傷。消費者C基於過失侵權告A經銷商求償，A經銷商在同一程序也起訴告B（impleader claim），結果還沒審判，A自己先跟C和解，C就撤回對A的告訴，這時候依照紐約州的法律，A就不能再跟起訴向B要求contribution。相同的情況，如果是B跟C先和解，A也不能告B要求contribution，但是A對C的責任數額也會減除B、C間約定和解金額、實際支付和解金額、與B應分擔責任金額，三者間最大的數額^{（註3）}，因為A或B有一個人已經脫離訴訟了，法院沒辦法認定A跟B個別要分攤多少比例的責任。但如果消費者C基於嚴格商品責任告A經銷商求償，A本來就可以回頭向B製造商要求百分之百的賠償，法院不用傷腦筋於怎麼分配A跟B怎麼分攤。所以A和解後，還是可以要求B全額償還（indemnity）。

因此，在契約裡面，如果想要indemnity就寫清楚是「indemnity」，盡量不要用別的字眼，例如「compensate」給自己找麻煩。

Indemnity跟guaranty與surety也有一些差別。Indemnity是我跟你說，你如果被告了，被判賠償第三人，我就替你繳錢。Indemnity是名義債務人跟實質債務人之間的約定，而且通常是關於侵權責任的承擔。而guaranty與surety則是我跟你說，他今天欠你錢如果不還，我替他還。或者我跟你說，他今天跟你簽契約做生意，他如果違約，我替他承擔責任，也就是當保證人。guaranty與surety是債權人與保證人之間的約定，而且通常是關於契約責任的承擔。Guaranty與surety的差別則在於guaranty還有先訴抗辯，surety沒有先訴抗辯權，也就是連帶保

3　N.Y. Gen. Oblig. Law § 15-108.

證（註4）。但這是英美普通法的用語，在國際擔保函的情境下，gua-rantee被當作是無先訴抗辯的。Guarantee跟guaranty常被混用。所以，在契約裡寫清楚，到底有沒有先訴抗辯權，這才是國際交易中的正途，否則打官司的時候，大家各搬一堆字典來吵，沒完沒了。

彌償的意涵有兩大構成部分，也就是提供辯護（defend）與填補損害及或各種費用支出（indemnify or hold harmless）。

以下我們提供實際的例句供讀者參考。

例句一：

7. INDEMNIFICATION AND INSURANCE（註5）

7.1 Indemnification by BIOMEDICA. BIOMEDICA agrees to indemnify, defend and hold MIP and its Affiliates and their respective directors, officers, employees and agents, harmless from and against any damages, claims, liabilities and expenses (including, but not limited to, reasonable attorney's fees) resulting from claims or suits ("General Claims Against MIP") arising out of (a) BIOMEDICAS' or a third party's use, handling or shipping of the Compound or Product (including in the event that MIP or MIP's designee makes shipping arrangements on behalf of BIOMEDICA), (b) BIOMEDICAS' breach of any of its material obligations, warranties or representations hereunder, (c) BIOMEDICAS' negligent acts or omissions or willful

---

4　參閱所引Pierce & Hefter著作第249，250頁。

5　美國證管會上市公司申報Supply Agreement，https://www.sec.gov/Archives/edgar/data/1340752/000119312510052762/dex105.htm，最後點閱日：August8，2023。

misconduct. Notwithstanding the foregoing, BIOMEDICA will not be required to indemnify, defend and hold MIP and its Affiliates and their respective directors, officers, employees and agents harmless from and against any General Claims Against MIP to the extent that such claims arise out of (i) MIP's breach of any of its obligations, warranties or representations hereunder; (ii) MIP's negligent acts or omissions or willful misconduct; (iii) any failure of MIP to supply (except to the extent labels and/or content thereof are provided by BIOMEDICA) or prepare for shipment of the Compound or Product in accordance with this Agreement, cGMPs or any other applicable laws, rules, regulations or other requirements of any applicable governmental entity; or (iv) any failure of MIP to supply Compound or Product consistent with the Specifications and requirements set forth herein.

　　7.2 Indemnification by MIP. MIP agrees to indemnify, defend and hold BIOMEDICA and its Affiliates and their respective directors, officers, employees and agents, harmless from and against any damages, claims, liabilities and expenses (including, but not limited to, reasonable attorney's fees) resulting from claims or suits ("General Claims Against BIOMEDICA") arising out of (a) MIP's supply (except to the extent that labels and/or content thereof is provided by BIOMEDICA) or preparation for shipment of the Compound or Product; (b) MIP's breach of any of its material obligations, warranties or representations hereunder; (c) MIP's negligent acts or omissions or willful misconduct; or (d) any failure of the

Compound or Product to meet the Specifications. Notwithstanding the foregoing, MIP will not be required to indemnify, defend and hold BIOMEDICA and its Affiliates and their respective directors, officers, employees and agents harmless from and against any General Claims Against BIOMEDICA to the extent that such claims arise out of (i) BIOMEDICAS' breach of any of its obligations, warranties or representations hereunder; (ii) BIOMEDICAS' negligent acts or omissions or willful misconduct; (iii) BIOMEDICAS' or third party's use, labeling, handling or shipment of the Compound or Product.

7.3 Conditions of Indemnification. A Party or any of its Affiliates or their respective employees or agents (the "Indemnitee") that intends to claim indemnification under this Article 6 shall promptly notify the other Party (the "Indemnitor") of any Liability in respect of which the Indemnitee intends to claim such indemnification reasonably promptly after the Indemnitee is aware thereof, and the Indemnitor shall assume the defense of any related third party action, suit or proceeding with counsel mutually satisfactory to the Parties; provided, however, that an Indemnitee shall have the right to retain its own counsel and participate in the defense thereof at its own cost and expense. Indemnity shall not apply to amounts paid in settlement of any claim, loss, damage or expense if such settlement is effected without the consent of the Indemnitor, which consent shall not be withheld unreasonably. The failure of an Indemnitee to deliver notice to the Indemnitor within a reasonable time after becoming aware of any such matter, if prejudicial to the Indemnitor's ability to defend such action, shall relieve the Indemnitor of any liability to the Indemnitee under this Article 6. The Indemnitee under this Article 6 and its employees and agents shall cooperate fully with the Indemnitor and its legal representatives in the investigation and defense of any matter covered by this indemnification.

7.4 Insurance. Each Party shall obtain and maintain insurance reasonably sufficient to cover its potential liability under this Agreement and shall provide evidence of such insurance to the other Party upon request.

7.5 Limitation of Damages. Neither Party nor its affiliates shall have any liability for any special, incidental, or consequential damages, including, but not limited to the loss of opportunity, revenue or profit, in connection with or arising out of this Agreement, even if it shall have been advised of the possibility of such damages.

第二節　智慧財產權條款

各種交易關係有時也會涉及智慧財產權的創設、授權或移轉,而需要特別約定相關的權利義務。例如僱人工作,或委任他人辦理事務,或共同開發研究,都可能產生各種智慧財產權。

智慧財產權的授權或移轉如果屬於主要履約內容,則應在雙方主要給付內容中約定清楚,但有時智慧財產的約定只是附隨性質的,那就寫在附隨義務條款組合中。

　　智慧財產權條款的內容通常包括以下四個構成部分：(1)確認雙方在交易或合作之前已有的智慧財產權，並不因為交易或合作而使權利歸屬發生變動；(2)界定因為雙方交易或合作而衍生的智慧財產權；(3)約定衍生智慧財產權的權利歸屬；(4)依據權利歸屬原則，約定配合之義務。第(1)個構成部分較簡易明瞭，暫不贅述。而其他構成部分在以下逐一補充說明。

1. 衍生智慧財產權的界定

　　界定因為雙方交易或合作而衍生的智慧財產權時，必然需要先對智慧財產權的分類體系有大致的理解，才能周全地約定。智慧財產權大致分為著作權、專利權、商標權、營業秘密，與基於競爭法的權利（例如不涉及商標的仿冒行為仍可主張權利）。各個法域的分類細節可能有所差異，但大體上是分成這幾類，也可以最後再加上其他類別以避免掛一漏萬。

　　約定衍生智慧財產權的權利歸屬時，需要特別說明，智慧財產權也具有類似於物權的性質，所以可再分成(1)所有權、(2)用益權與(3)擔保權。因此，約定權利歸屬時，需特別留意按照所有權、用益權與處分擔保權，三種不同的權利具體約定。

　　著作權、專利權、商標權還有更細節的內容分類，在約定權利歸屬時，必須掌握內容分類。就像一棟透天，可以整棟出租，也可以在二樓以上分割成多個套房供住宅使用，與一樓商業店面分別出租，以獲得最大利益。

　　著作權首先可以區分成著作人格權（moral rights），與著作財產權。人格權是標示誰是著作人，財產權則是對著作物為使用收益的權

利。

　　著作權內容的分割方法可以參考法律的分類，臺灣著作權法第二條第一項第五款到第十五款列了十一種權利內涵（重製、公開口述、公開播送、公開上映、公開演出、公開傳輸、改作、散佈、公開展示、發行、公開發表）[註6]，當中有些只適用於特定著作物，例如公開演出，適用於劇本、播送適用於影音著作物。重疊再組合的部分。

　　我們可以在前述法律分類的基礎上再按實際需求細分授權內容。舉例來說，我們可以看到電影或電視劇的著作權授權有極多的態樣，可以分成網路傳送、電視播映、分區域、分語言、甚至分高清版與非高清版、或只用其中幾分鐘，幾秒鐘的片段進行授權。而實體的影音DVD既可以分成出售版或出租版，還可以再細分成家用版或公播版。

2. 智慧財產權的權利歸屬

　　必須先掌握前述個別智慧財產權利的分類，才能具體約定智慧財產

6　「五、重製：指以印刷、複印、錄音、錄影、攝影、筆錄或其他方法直接、間接、永久或暫時之重複製作。於劇本、音樂著作或其他類似著作演出或播送時予以錄音或錄影；或依建築設計圖或建築模型建造建築物者，亦屬之。六、公開口述：指以言詞或其他方法向公眾傳達著作內容。七、公開播送：指基於公眾直接收聽或收視為目的，以有線電、無線電或其他器材之廣播系統傳送訊息之方法，藉聲音或影像，向公眾傳達著作內容。由原播送人以外之人，以有線電、無線電或其他器材之廣播系統傳送訊息之方法，將原播送之聲音或影像向公眾傳達者，亦屬之。八、公開上映：指以單一或多數視聽機或其他傳送影像之方法於同一時間向現場或現場以外一定場所之公眾傳達著作內容。九、公開演出：指以演技、舞蹈、歌唱、彈奏樂器或其他方法向現場之公眾傳達著作內容。以擴音器或其他器材，將原播送之聲音或影像向公眾傳達者，亦屬之。十、公開傳輸：指以有線電、無線電之網路或其他通訊方法，藉聲音或影像向公眾提供或傳達著作內容，包括使公眾得於其各自選定之時間或地點，以上述方法接收著作內容。十一、改作：指以翻譯、編曲、改寫、拍攝影片或其他方法就原著作另為創作。十二、散布：指不問有償或無償，將著作之原件或重製物提供公眾交易或流通。十三、公開展示：指向公眾展示著作內容。十四、發行：指權利人散布能滿足公眾合理需要之重製物。十五、公開發表：指權利人以發行、播送、上映、口述、演出、展示或其他方法向公眾公開提示著作內容。」

權如何歸屬。就所有權的部分，契約條款除了確認誰單獨具有所有權，或各方共有的比例之外，也需要約定誰有權處分所有權。用益權的部分，則是要約定可否個別自行利用、可否授權他人使用、授權的利益如何分配等等。擔保權則是可否為他人之利益以智慧財產設定擔保，以及設定擔保需要的共有人合意程序。常常看到有些契約條款只簡略地約定衍生智慧財產權，由契約當事人共有，有些約定了共有比例，有些未約定比例，但未特別約定對外授權的權限歸屬，則依法解釋是否對外授權需要全體共有人同意呢？若是如此，可能極為不便。智財權對外授權可能需要有更便利的約定，可能某一方有授權專業人力，能較妥適地處理。例如合作拍攝電影或電視劇，可能約定由出品人一方負責對外授權工作，取得的權利金，各出品人按照比例分潤，也可能約定多個出品人都可以對外授權，談成授權的出品人可以多分一些，鼓勵各方當事人努力對外協商授權。

3. 依據權利歸屬原則，約定配合之義務

　　專利權需要申請、商標也需要登記，著作權雖然並未強制登記，但辦理登記仍有一些方便證明的好處，例如電影、電視劇對外授權時，就需要向相關公協會申辦著作權證。所以在契約的智慧財產權條款，也可能需要約定各方配合登記提供簽署特定文件的義務。

　　以下我們提供契約條款實例供讀者參考。

例句二：

7. INTELLECTUAL PROPERTY OWNERSHIP. [註7]

7.1 Background Technology and Intellectual Property Rights. Each party shall retain all rights, title, and interest in and to its Background Technology and all Intellectual Property Rights therein that have arisen by the Effective Date. "Background Technology" consists of each party's or its licensors' technology, materials, ideas, know-how, inventions, approaches, software, hardware, microelectronics, designs, concepts, techniques, processes, data, tools, services, instruments, templates, methodologies, algorithms, documentation and any other knowledge, and any derivation thereof or enhancements or modifications thereto. "Intellectual Property" or "Intellectual Property Rights" collectively means any and all patents (including reissues, divisions, continuations and extensions thereof), patent registrations, patent applications, database rights, utility models, business processes, trademarks, service marks, trade secrets, know-how, trade names, registered or unregistered designs, mask works, copyrights, moral rights, industrial rights, or any application therefor and any other form of proprietary protection, which arise or are enforceable under the laws of the United States, the European Union, Japan, Taiwan, any other jurisdiction or any multilateral, bilateral or other treaty regime.

7　美國證管會上市公司申報 Development Agreement，https://www.sec.gov/Archives/edgar/data/153 7054/000119312512191562/d267959dex1013.htm, 最後點閱日 August 10，2023。

7.2 Project IP. Any and all inventions, improvements, technology, developments, innovations, ideas, know-how, approaches, software, hardware, designs, concepts, techniques, processes, data, tools, templates, methodologies, algorithms, documentation and any other Intellectual Property which is developed by QUALCOMM or jointly by QUALCOMM and AirCell during the Term and pursuant to this Agreement for any Deliverable is "Project Work Product." QUALCOMM shall be the sole and exclusive owner of all Intellectual Property Rights that are not AirCell's Background Technology or Background Intellectual Property Rights in and to the Project Work Product ("Project IP"). QUALCOMM shall not own any Intellectual Property Rights developed independently by AirCell. QUALCOMM shall not own any products, technology or any Intellectual Property of AirCell under this Agreement and the Statement of Work.

7.3 Grant of License to AirCell Background Technology. Subject to the terms of this Agreement, AirCell hereby grants to QUALCOMM a non-exclusive, worldwide, royalty-free, and fully paid-up right and license (with no right to sublicense) to use the AirCell Background Technology and AirCell Intellectual Property rights solely during the Term and for the sole and limited purpose of QUALCOMM's performance of its obligations hereunder.

7.4 Grant of License to Deliverables. Subject to the limitations set forth in Section 7.5 below, for any and all Deliverables provided by QUALCOMM to AirCell under this

Agreement, AirCell may use the Deliverables solely in connection with the testing and development of wireless
communications equipment that incorporates a QUALCOMM
integrated circuit. AirCell shall have no right to alter, modify,
translate or adapt the Deliverables or create derivative works
thereof except as expressly provided in Section 7.5, nor shall
AirCell have the right to assign, sublicense, transfer or otherwise
provide the Deliverables to any third party (except for the
provision of the Deliverables to AirCell's consultants, advisors
and other agents (the "AirCell Consultants") who have a need to
access the Deliverables to fulfill the purpose of this Agreement;
provided that such AirCell Consultants have agreed with AirCell
to treat such Deliverables in a manner that is consistent with the
obligations of AirCell in this Agreement), or as set forth in the
last sentence of this Section 7.4, and in the Supply Agreement
and/or any Design Transfer Agreement. AirCell shall be
responsible for any improper use by the AirCell Consultants of
such Deliverables. Except as expressly permitted above, AirCell
shall not use the Deliverables for any other purpose, without the
prior written authorization of QUALCOMM. QUALCOMM
covenants and agrees that in the Supply Agreement and/or a
Design Transfer Agreement, it will grant to AirCell all necessary
rights and licenses to use, copy, display, modify, reproduce,
manufacture, have manufactured, market, sell and distribute the
Deliverables, or parts thereof, as a component of AirCell
products.

7.5 Grant of Software License. QUALCOMM hereby grants

to AirCell a non-exclusive, non-transferable, revocable license under QUALCOMM's copyrights in the QUALCOMM Background Technology and the Project IP to use the software supplied hereunder by QUALCOMM (the "Software") solely in conjunction with the Deliverables provided hereunder and subject to the terms and conditions of this Agreement . In addition, AirCell shall have the right to have one or more BTS vendors, as selected by AirCell, incorporate the CSM Software (that is part of the Deliverables) into a BTS (an "Enabled BTS") and shall be permitted to resell, deploy and use such Enabled BTS on a worldwide basis. The BTS vendor(s) of AirCell's choice shall have appropriate QUALCOMM licenses; however they will not be required to pay any additional license fee for the CSM Software (ATG features). The selected BTS vendor(s) shall have the right with AirCell's prior written permission and upon written notice to QUALCOMM to sell and deploy the Enabled BTS to any service provider worldwide. Except as expressly provided in this Section 7.5, AirCell warrants and agrees that AirCell shall not, without the prior written consent of QUALCOMM, (i) alter, modify, translate, or adapt any Software or create any derivative works based thereon; (ii) except as necessary to install or load the Software in the Deliverables, copy any Software; (iii) assign, sublicense, resell or otherwise transfer the Software in whole or in part to any unauthorized third parry; (iv) transfer Software except in conjunction with the transfer of the product in which the Software is imbedded or contained; (v) use the Software except as specifically contemplated in this Agreement; (vi) decompile, reverse assemble, translate or otherwise reduce the Software or any portion thereof to human-perceivable form; (vii) combine or merge any portion of the Software with any other software; (viii) disclose the Software to any third party (other than in connection with the permitted uses of the Deliverables); or (ix) incorporate, link, distribute or use (1) the Software, or (2) any software, products, documentation, content or other materials developed using the Software, with any code or software licensed under the GNU General Public License ("GPL"), LGPL, Mozilla, or any other open source license, in any manner that could cause or could be interpreted or asserted to cause the Software or other QUALCOMM software (or any modifications thereto) to become subject to the terms of the GPL, LGPL, Mozilla or such other open source license. The entire right, title and interest in the Software shall remain with QUALCOMM, and AirCell shall not remove any copyright notices or other legends from the Software or any accompanying documentation. Nothing herein shall be construed as the sale of any Software to AirCell. Nothing herein shall be deemed to grant any right to AirCell under any of QUALCOMM's patents. This Agreement shall not modify or abrogate AirCell's obligations under any other agreement with QUALCOMM. Neither the supply of any Deliverables nor the license of any Software, nor any provision of this Agreement shall be construed to grant to AirCell either expressly, by implication or by way of estoppel, any license under any patents or other intellectual property rights of QUALCOMM covering or relating to any other product or invention, or any combination of any Deliverable with any other product, except as expressly set forth in this Agreement.

7.6 Ownership of Third Party Materials. AirCell may disclose or provide to QUALCOMM certain Intellectual Property which is owned by third parties and licensed to AirCell (the "Third Party Intellectual Property"). Notwithstanding anything express or implied in this Agreement, such third party owners shall retain all rights, title and interest in and to such Third Party Intellectual Property including all underlying Intellectual Property Rights. Any such disclosure of information related to Third Party Intellectual Property shall be subject to Section 9 (Confidential Information) of this Agreement.

7.7 Further Assistance. The parties agree to execute all applications, assignments or other documents of any kind and take all other legally necessary steps under the law of any applicable jurisdiction including the United States-or any applicable treaty regime, at the expense of the requesting party, in order to apply for, obtain, protect, perfect or enforce the requesting party's rights, title, and interest in the requesting party's Intellectual Property as specified herein.

7.8 Reservation of Rights. Except as expressly provided in this Agreement, neither party conveys to the other party any Intellectual Property Rights. Neither the delivery of any QUALCOMM or AirCell Background Technology, the Project IP, or other Intellectual Property, nor any provision of this Agreement shall be construed to grant to either party, either expressly, by implication or otherwise, any license under any Intellectual Property Rights of the other party other than the limited licenses granted in Section 7.3, Section 7.4 and Section 7.5.

契約的轉折部分，
與情事變更相關條款

　　介紹了啟始部分（標題、當事人介紹、敘言、合意文句），以及承接主軸（效期、核心事項條款、保密、競業、彌償、智慧財產權等各類附隨義務）之後，我們接著要說明契約的轉折部分，主要包括情事變更相關條款、賠償條款、終止解除等散場條款、以及紛爭解決條款。

第一節　意外跟明天誰先來？

案例一[註1]

　　被告英國商人在1914年與英國的原告簽約，約定銷售芬蘭的木材給原告。契約並未約定戰爭條款或不可抗力條款，或其他有關暫停履約的條款。原告不知道通常做法，是從芬蘭港口直接運木頭到英國，原告也不知道賣方英國商人並沒有木材庫存。第一次世界大戰爆發，被告主張因為戰爭無法交付芬蘭木材，法院認定被告仍無法免責。因為被告可以交付英國木材作為替代，雖然英國木材砍伐的費用較高，運送費用也較高。

　　所謂計劃趕不上變化，契約履行過程中可能出現特殊事件，造成無法履約，或者按照原定條件履約會變得不太公平的情況，如何處理這些狀況呢？

　　不同國家對於履約過程中出現的意外變化有不同的規範方法，例如臺灣民法第227條之2第一項規定「契約成立後，情事變更，非當時所得預料，而依其原有效果顯失公平者，當事人得聲請法院增、減其給付或變更其他原有之效果。」

1　Blackburn Bobbin Co., Ltd v. T.W. Allen & Sons, Ltd., 1 K.B. 540, 551.

臺灣民法第230條也規定「因不可歸責於債務人之事由，致未為給付者，債務人不負遲延責任。」僅提及「不可歸責於債務人之事由」，並未對不可抗力做更詳細的說明。若履約已永久地不能，固然可依臺灣民法第255條的給付不能處理。但若只是暫時的履約不能，在履約障礙事由持續作用的期間內，除去免除遲延責任之外，締約當事人間仍有其他的權義關係必須更細膩地加以安排。

聯合國國際貨物銷售合同公約第79條第1項規定^(註2)，如果義務人能證明之所以未履行義務，是因為發生了超出義務人控制能力的障礙事由，而且對於此種障礙，在訂定契約時無法合理預期，或者雖可以預期，但無法迴避或克服障礙本身或其後續影響，則義務人不必承擔不履行的責任。同條第2項規定，如果義務人委託第三人履行，則義務人或第三人遇到第一項的障礙，未履行義務時，也可以免責。第3項則規定，只有在障礙存在的期間內，才有免責效果。第4項規定，義務人因障礙事由未履行義務時，必須通知契約的權利人，如果權利人在未依約履行的事實發生之後，一段合理期間內仍未收到通知，則義務人必須為了沒有通知到權利人，使權利人受損害，承擔責任。第5項則規定，權利人遇到義務人因為不可抗力事由不履約時，不能主張損害賠償，但可以依照公約其他規定主張權利。也就是說，權利人可以依照公約第四49條第1項第1款規定，解除契約。如果權利人沒有解除契約，當然在障礙事由消滅之後，義務人仍必須繼續履行^(註3)。

國際統一私法協會（International Institute for the Unification of Private Law, UNIDROIT）制訂的2010年版國際商事合同通則Prin-

---

2　聯合國國際貨物銷售合同公約第79條第1項，http://www.uncitral.org/pdf/chinese/texts/sales/cisg/V1056996-CISG-c.pdf，最後點閱日：July 9，2011。

3　Alejandro M. Garro, *Force Majeure And CISG Article 79: Competing Approaches And Some Drafting Advice* in *Drafting Contracts Under The CISG* 380 (Harry M. Flechtner et al. ed., Oxford 2008).

ciples of International Commercial Contracts）第7.1.7條基本上援用聯合國貨物銷售合同公約第79條的規定，但刪除有關義務人委託第三人履約時，第三人面對所造成無法履行的情況，並特別說明權利人仍可終止契約、暫停自己的給付，或請求金錢給付的利息[註4]。

中國民法典第563條規定第一項規定「因不可抗力致使不能實現合同目的，」當事人可以解除合同。

中國民法典第590條規定「1.當事人一方因不可抗力不能履行合同的，根據不可抗力的影響，部分或者全部免除責任，但是法律另有規定的除外。因不可抗力不能履行合同的，應當及時通知對方，以減輕可能給對方造成的損失，並應當在合理期限內提供證明。2.當事人遲延履行後發生不可抗力的，不免除其違約責任。」此種規定有些類似聯合國國際貨物合同銷售公約，但民法典第180條對不可抗力的定義似乎較為嚴苛，必須「不能預見、不能避免、不能克服」。若是契約條款中舉例「火災、洪水」是否已預見，而這些例子不能主張不可抗力？而聯合國國際貨物合同銷售公約規定，則是不能預見，或雖能預見但不能避免，或不能克服障礙本身或其影響。

由以上的規範，我們可以整理出三種相關而近似的概念，也就是不可抗力（force majeure）、艱困情況（hardship），與情事變更（change of circumstances）。不可抗力原本指的是發生某種意外狀況，造成履約在現實上不可能，當事人可能約定契約條款來處理這些問題，或者任由法律以預定規則，來解除契約或免除遲延責任或不履行責任。艱困情況指的是發生某種情況，若依照契約原來的約定履行，會使雙方權義顯失平衡。而情事變更的原因與後果與艱困條款相近，但情事

---

4 UNIDROIT, Principles of International Commercial Contracts 2010 §7.1.7, http://www.unidroit.org/english/principles/contracts/principles2010/blackletter2010-english.pdf, latest accessed on 2011/7/9.

變更原則，通常是由法院來增減或調整給付內容，使失衡的權義再次平衡。而艱困情況，通常是契約當事人自己明訂條款約定處理方式，而不適用法律規定。

Force Majeure是出自於法國的拿破崙民法典，而在西歐地區廣泛採用，但具體定義與適用要件可能各不相同，例如不可抗力事件是否必須是不可預見的？以及債務人委託第三人履約時，第三人遇到不可抗力事件，是否能讓義務人免責？(註5)。

英美法雖無Force Majeure的概念，但案例法也有其他相近的機制，例如締約目的受挫（frustration）、履行不可能的抗辯（impossibility）、以及商務上履約不實際的抗辯（commercial impracticability），美國統一商法典則將商務上履約不實際的抗辯成文化。

美國契約法第二次整編第265條表示，契約訂定之後，一方當事人主要締約目的，因為不可歸責之事件而嚴重受挫時，若該事件之不發生是契約訂定之時的基本假定，則除非契約另有約定或個案有其他特殊情形外，目的受挫一方履約責任即消滅。第266條第2項表示，締約時，當事人一方的主要締約目的因為不可歸責，且無從得知的情事而嚴重受挫時，若該事件之不發生，是契約訂時的基本假定，則除非契約另有約定或個案有其他特殊情形外，目的受挫一方履約責任即消滅。此即締約目的受挫抗辯之意含。

而普通法上的履行不可能抗辯對於債務人極為嚴苛，法院通常認定當事人締約時，就應該承擔失去履約能力的風險，所以必須發生的事件，確確實實造成不可能履約的情況，才有可能免責，例如在提供屬人性較強的服務契約中，義務人死亡，新通過的法令讓契約的履行變成違

5　Richard Christou, *International Agency, Distribution and Licensing Agreements* 42-43 (4th ed. Sweet & Maxwell 2003).

法，或者契約標的滅失。有時候即使情況極為惡劣，履約極其困難，法院仍不免除義務人責任，例如案例一的木材買賣契約糾紛。

因此，為了調和履行不可能此一抗辯的嚴苛要求，普通法又發展出商務上履約不實際的抗辯。但具體操作上，各法院有不同見解。有的法院認定障礙事件能否預見是關鍵要素，有的法院則分析各方當事人中，哪一方較適於承擔風險，或採取保險行為加以轉嫁。

美國統一商法典第2-615條規定，當事人在締約時以某種特殊事件不會發生為基本假定，但該特殊事件卻真的發生了，從而造成商品出賣人實際上難以履約，而遲延交貨、或未交貨，或者為了遵守國內政府或國外政府的法規、命令，而遲延交貨、或未交貨；即使之後該法令被證明是無效的，出賣人都可以免責。但是出賣人如果還能履行契約中的一部分，就必須在各個客戶間合理地配貨運送。而且出賣人也必須適時地通知賣方此種履約不實際的情事。

統一商法典僅適用於貨物買賣，而且只適用在賣方未依約交貨責任的免除。在運用上，法院仍會參照普通法的規則，從而運用結果同樣難以預測。[註6]

此外，英美法上履約目的受挫、履約不可能、或履約不實際等抗辯事由的效果，與不可抗力的效果亦有些出入。不可抗力的效果通常在於暫時停止履約，且不課停止履約一方違約責任，當然一段時間過後，若履約障礙未消除，則當事人可以終止或解除契約。然則履約目的受挫、履約不可能、或履約不實際等抗辯事由的效果，則是讓義務人可以主張直接免除責任，從而使契約消滅；當然對方如果已經做出部分的給付，就看給付多少可以依照準契約（quasi-contract）的理論請求部分對價。

---

6 Nancy M. Persechino, *Force Majeure* in *Negotiating And Drafting Contract Boilerplate* 321-325 (Tina L. Stark et al. ed., ALM Publishing 2003).

　　因此在個別交易中，宜考量哪些情況安排成不可抗力事由，雙方必須稍做等待，等確定障礙無法排除，才解除或終止契約。而哪些情況一旦發生可以直接列為終止或解除契約的事由，而不需要不可抗力條款的等待期。

　　從以上的簡要說明，我們可以瞭解各國的相關規範有很大的差異，因此締約當事人宜參照各國的法規，訂出符合個別交易需求的不可抗力條款，否則等到發生糾紛，法院或仲裁庭依照準據法，做出的裁判或判斷可能超乎締約當事人的預期。

第二節　情事變更相關條款的設計

　　為因應履約過程，可能發生的意外狀況，而設計的契約條款就是不可抗力條款（Force Majeure）、艱困條款（Hardship）與履約內容自動調整條款。這三種條款的內容都是 (1)發生意料之外的情況，(2)造成履約障礙，(3)雙方約定處理方法。只是這三個構成部分的細節內容可能有所差異。履約內容自動調整條款在處理方法上先預定增加價格、減少給付，或變更給付內容等方式來因應問題，如果解決不了，由第三人裁斷，或解除或終止契約。而不可抗力與艱困條款則未必預定自動調整履約內容，可能約定先等候一段期間或直接約定終止解除契約。

　　不可抗力條款與艱困條款的差異在於，不可抗力事件讓當事人無法履約，但艱困事件或重大情事變更，只是讓履約變得更費成本，從而對履約方造成重大不利益，但並未造成確實不能履約的情況。法國與比利時採取較嚴格的imprévision理論，要求此類艱困事件，必須造成當事人間權利義務失衡。但其他法域可能接受情況變化，造成當事人一方締約目的無法達成的情況，例如荷蘭新民法典（Nieuw Burgerlijk Wetboek）§6-259即將此類特殊事件，定義為任何依照理性與衡平的標準，他方無法期待按契約原來內容履行的情況。（註7）此外，事件的發生時點，能否預見、應否預見、實際上是否被預見，因特殊事件而蒙受不利益的一方，對於此類事件有無能力加以控管，以及當事人間對於此類事件的風險是否已有其他安排等等（註8），都是界定艱困條款觸發事件時的考量點。臺灣民法第227-2條的情事變更原則亦屬類似的概念。

所以我們將這三種條款整併起來探討，在實際撰擬契約時，可視個別情況，名之為不可抗力條款，或情事變更條款，或自動調整履約內容條款。

1. 發生意料之外的情況

從第一節說明的各法域法律規定，我們可以整理出意外狀況的本質，就是(1)超出義務人控制的事件，而還有幾個可能附加的要件；包括(2)不能預見、不能避免、不能克服、或當事人基本預設不會發生；(3)狀況的發生，可否歸責於義務人，除了義務人之外，義務人的代理人、或履行輔助人所遇到的不可抗力事件，是否算入不可抗力事件最好也約定清楚；(4)而如果不以不能預見作為要件，還可能舉例，常見的不可抗力事件有生產工具或原料無法取得、地震、水災、颱風、海嘯、火災、戰爭、疫情、罷工、圍廠、船隻捕獲、法令變更、政府封鎖、民眾示威、政府法律政策變更等等。

---

7   Marcel Fontaine & Fillip De Ly, *Drafting International Contracts: An Analysis of Contract Clauses* 454 (Martinus Nijhoff 2009).

8   International Institute for the Unification of Private Law, *UNIDROIT Principles of International Commercial Contracts 2010*, Article 6.2.2, http://www.unidroit.org/english/principles/contracts/main.htm, latest accessed on 2011-09-01. 亦可參閱陳自強著，《契約法中艱困法則之比較研究研究成果報告（精簡版）》（行政院國家科學委員會專題研究計劃，2007年10月29日），5-8頁。

2. 造成履約障礙

意外造成的履約障礙,可能約定是暫時或永久不能履行(不可抗力)、可能是依原定方式履約會造成權義失衡而不實際(艱困條款或情事變更條款)。

3. 應變方法

應變方法主要內容大致說就是三種等、調整給付內容、解除終止契約。等待可能需要明定等多久,如果順利,就要回復履約,如果等候時間終了,障礙仍未排,還是要做其他調整。而調整給付內容可能是設定自動調整的公式(註9),也可能是約定由第三方(法院、仲裁庭、專家小組)來調整。而除了終極處理方法之外,還可能約定發生意外狀況時,有儘早通知的義務。

如前所述,不可抗力事件在不同的法域,根據不同的理論基礎,例如締約目的受挫,可能有不同的效果,因此當事人如果不約定清楚,彼此之間白紙黑字約定的處理方式是不可抗力事件的唯一處理方式,一旦發生事情,很可能有一方會依據這些理論做爭執,結果將難以預測,因此宜明文約定當事人間合意的方式是唯一有效的處理方案。

以下提供不可抗力條款的例句供讀者參考:

---

9 例如按照消費者物價指數調整租金。美國統一商法典2-615條第1項(b)款還規定,若意外情況只影響商品賣方的部分產能,賣方必須在各個買家之間公平分派商品,不能只交貨給某一家或幾家客戶,而完全不交貨給其他客戶。再次強調,統一商法典之適用於B2B的商品銷售,不適用於B2C或C2C的銷售。

例句一：（註10）

Section X.02 Suspension of Performance. Subject to the provisions of Sections X.03 and X. 04, if a *Force Majeure* Event occurs, the Nonperforming Party is excused from

(a) whatever performances is prevented by the *Force Majeure* Event to the extent so prevented (a "Suspension of Performance"); and

(b) satisfying whatever conditions precedent to the Performing Party's obligations that cannot be satisfied, to the extent they cannot be satisfied (a "Suspension of Performance")

Section X.03. Obligations of the Nonperforming Party.

(a) *Written Reports.*

(i) No later than two working days after becoming aware of the occurrence of a *Force Majeure* Event, the Nonperforming Party shall furnish the Performing Party a written report describing the particulars of the occurrence, including an estimate of its expected duration and probable impact on the performance of the Nonperforming Party's obligations under this Agreement.

(ii) During the continuation of the *Force Majeure* Event, the Nonperforming Party shall furnish timely, regular written reports, updating the information required by Section X.03(a)(i) and providing any other information

---

10 參酌改編 Persechino 前揭文第 355-357 頁。

that the Performing Party reasonably requests.

(b) *Other Obligations*. During the continuation of the *Force Majeure* Event, the Nonperforming Party shall

(i) exercise commercially reasonable efforts to mitigate damages to the Performing Party;

(ii) exercise commercially reasonable due diligence to overcome the *Force Majeure* Event;

(iii) to the extent it is able, continue to perform its obligations under this Agreement; and

(iv) cause the Suspension of Performance to be of no greater scope and no longer duration than the *Force Majeure* Event requires.

Sections X.04. Conditions Precedent.

(a) *Conditions Precedent to Initial Suspension of Performance*. The Nonperforming Party's performance of the covenant set forth in Section X.03(a)(i) is a condition precedent to its initial Suspension of Performance. If the covenant is performed, the Suspension of Performance is deemed to have commenced on the date the *Force Majeure* Event occurred.

(b) *Conditions Precedent to Continued Suspension of Performance*. During the continuation of the *Force Majeure* Event, the Nonperforming Party's performance of the covenants set forth in Section X.03(a)(ii) and Section X.03(b) are conditions precedent to its continued Suspension of Performance.

Section X.05. Resumption of Performance. When the Nonperforming Party is able to

(a) resume performance of its obligations under this

Agreement, or

(b) satisfy the conditions precedent to the Performing Party's obligations.

it shall immediately give the Performing Party written notice to that effect and shall resume performance under this Agreement no latter than two working days after the notice is delivered.

Section X.06. Termination. If the Suspension of Performance continues for a period of more than twelve consecutive months as a result of a *Force Majeure* Event, either party is entitled to terminate this Agreement by giving a notice to the other party pursuant to the notice provisions of this Agreement.

Section X.07. Exclusive Remedy. The relief offered by this *Force Majeure* provision is the exclusive remedy available to the Nonperforming Party with respect to a *Force Majeure Event*, and the parties waive all the other protection or defense offered by any other applicable laws or regulations.

以下也提供國際商會的不可抗力示範條款。^{（註11）}

例句二：

A party is not liable for a failure to perform if he can prove that: (1) the failure was due to an impediment beyond his

---

11 ICC Publication No.421, "Force Majeure and Hardship"(1985)轉引自 Larry A. DiMatteo, *The Law Of International Contracting* 57 (Kluwer Law International 2000).

control, (2) he could not have reasonably foreseen the impediment at the time of contract formation, and (3) he could not have reasonably avoided or overcome its effects. An impediment includes but is not limited to :

　　(a) war, hostilities, and acts of piracy,

　　(b) natural disasters,

　　(c) explosions, fires, and destruction of machinery,

　　(d) boycotts, strikes, lock-outs, and work stoppages which occur in the enterprise of the party seeking relief,

　　(e) acts of authority.

　　A party seeking relief pursuant to this clause shall give notice as soon as practicable.

　　Failure to give timely notice makes the party liable in damages for losses that could otherwise have been avoided. Notice shall also be given when the impediment ceases.

　　The relief granted under this clause is a postponement of the time for performance for such period as may be reasonable. The nonbreaching party may also suspend his own performance. Either party may terminate the contract in the event that the impediment persists for a period of ＿ days. Upon the termination of the contract, each party may retain whatever he has received, but must account to the other party for any unjust enrichment.

　　以下提供艱困條款的例句供讀者參酌：

例句三： （註12）

Article 26 Hardship

26.1 If events occur which have not been contemplated by the Parties and which fundamentally alter the equilibrium of the present Agreement, thereby placing an excessive burden on one of the Parties in the performance of its contractual obligations, that Party shall be entitled to request revision of this Agreement.

26.2 The request for revision shall be addressed to the other Parties and the Management Committee. It shall indicate the grounds on which it is based.

26.3 In response to such a request, the Parties shall consult with a view to revising the Agreement on an equitable basis, so that no Party suffers excessive prejudice or burden.

26.4 If the Parties fail to reach agreement on the requested revision, any Party may resort to the proceedings provided in Article 32.2 and 32.3, and to arbitration pursuant to Article 32.4 to 32.7. The Arbitral Tribunal shall have the power to make any revision to this Agreement that it finds just and equitable in the circumstances.

---

12 International Trade Centre UNCTAD/WTO, Contractual Joint Venture Model Agreement (three parties or more), http://www.jurisint.org/doc/orig/con/en/2004/2004jiconen3/2004jiconen3.pdf, accessed on 2011/11/14.

違約救濟條款

第一節　違約救濟（Remedies）條款的寫法

違約救濟是由違約情事與對應的救濟內容組成的。違約救濟其實也是一種給付義務，跟一般給付義務的差別在於違約救濟，可以說是第二次的給付義務。因為義務人違反了第一次的給付義務，所以構成第二次的給付義務。

違約救濟條款有幾種設計方法。第一種方法是分散式的，也就是將個別的違約情事與對應的救濟內容，整合成一個條件式的給付義務句，寫在第一次給付義務之後。例如：

例句一：

Rent Payment. During the Term, the Tenant shall pay the Rent of the next month to the Landlord on or before 30th day of each month. If the Tenant fails to pay the Rent on time, the Tenant shall pay the Landlord additional 1% of the Rent as Overdue Charge per day since the due day through actual payment day.

第二種方法是集中式的，即是按照救濟內容的類型，將適用相同救濟內容的違約事項列在一起。例如：

例句二：

Acceleration. In any of the following event, the Lender may declare all the installments and interests the Borrower has not paid yet immediately due:

　　(a) The Borrower fails to pay any installments under this Agreement.

　　(b) The Borrower defaults under any other contract, the Lender may declare such default constitutes a Default in this Agreement.

　　(c) The Borrower defaults under any other contract and all the obligations not performed by the Borrower are declared immediately due.

　　如果一份契約在執行上能夠發生的違約情況不多，在說明個別給付義務的時候，附帶說明違約救濟，這種分散式的作法較為簡潔。如果適用相同救濟方法的違約情況很多，將各種可能的違約情事集中在一起，像十誡那樣列出來，然後再說明違約的救濟效果，此種集中式的作法較為一目瞭然。

第二節　可能的違約事件

　　違約事件有兩個主要的來源，一個來源當然是契約裡的約定事項。但契約之外的事項也有可能構成違約。例如，在貸款契約中，放款人經常會要求加入「連鎖違約條款」（cross default）、與「連鎖加速條款」（cross acceleration）以擴大違約事件的範圍。

　　連鎖違約條款指的是義務人，在別的契約中一旦發生違約，權利人也可以宣告義務人在本契約中也構成違約，所以按照違約救濟條款處理。這種條款是借款人怕自己變成最後一隻老鼠，如果債務人已經對別

人跳票了,還要等到自己的債權給付期限屆至再主張權利,就來不及了。連鎖違約條款範例如下:

> ## 例句三:
>
> If the Borrower defaults under any other contract, the Lender may declare such default constitutes a Default in this Agreement.

連鎖加速條款指的,則是義務人所簽訂的另一份契約中有分期給付約定,該契約的權利人因為義務人一期未給付,而宣告所有期限尚未屆至的給付義務一次到期,此時,本契約的權利人也宣告義務人,在本契約中構成違約。所以相較於連鎖違約條款來說,連鎖加速條款對義務人較寬容,必須義務人在別的契約違約,而且被宣告適用加速條款,使未到期的給付也立即到期,才會在本契約中被宣告構成違約。連鎖加速條款範例如下:

> ## 例句四:
>
> If the Borrower defaults under any other contract and all the obligations not performed by the Borrower are declared immediately due, the Lender may declare such event constitutes a Default in this Agreement.

至於契約約定事項可能產生的違約事件則較容易辨識,例如遲延交貨、遲延付款、工期遲誤、交貨品質不良、違反聲明與擔保內容,以及

違反其他各種給付義務^{（註1）}。

　　此外，除了正面列出違約事件外，也可以列出非屬違約的事件，以排除責任。這種做法類似於在聲明擔保條款中，列出免責聲明事項。當然，在撰擬契約時，特定事項究竟是列為免責聲明，還是放在違約條款中列為非違約事件，要看個別情況斟酌決定。某些法域可能規定有些責任不能預先約定排除，也需注意。

第三節　重大違約與輕微違約

　　在美國，違約有分重大違約，跟輕微的違約，兩者的效果不同。輕微的違約指的是，即使有違約，但權利人已經取得主要的履行利益。例如不嚴重的遲交、或品質、數量上有小幅度的落差。對於輕微的違約，權利人仍然必須履行自己的給付義務，但可以針對違約部分要求義務人賠償損害。

　　如果是重大違約，權利人可以結束契約關係，所以如果權利人也有給付義務，就可以不履行了，也可以立即要求義務人賠償整個契約的損害。

　　不過在跨國交易中，此種重大違約，與輕微違約效果上的差別或許不能放諸四海皆準，當事人自己約定清楚重大違約，與輕微違約的具體效果差異（例如可以終止、解除契約）較為妥當。

1　關於各類違約事件可參考 Marcel Fontaine & Fillip De Ly, *Drafting International Contracts: An Analysis of Contract Clauses* 302-328 (Martinus Nijhoff 2009)。

第四節　違約的救濟內容

發生違約情況時，各個國家都有法律或判例法可以主張請求救濟，但是各個國家的規定都不同，所以在跨國交易中，當事人可能希望自己講清楚，免得到時候大吃一驚，怎麼法律是這樣規定的。

最直接的救濟內容，就是強制履行（specific performance）、也就是請求法院將當事人要的標的直接進行移轉，或者變更給付內容來取代原給付，如果有多項標的，而現實上已無法獲得每一項標的，也可能以減價收受，終止或解除契約、行使加速條款（acceleration）、處分即到期（due-on-sale）、損害賠償，此外還可能有擔保物的拍賣換價，或作價收受（此項救濟方法在某些法域可能受到消保法、擔保交易法或破產法的限制），以及衡平法上的救濟，例如禁制處分（injuction）[註2]，甚至也可能透過不當得利返還、擬制信託、衡平擔保、強制權利移轉等方式進行救濟。強制履行雖然必須符合法律要件，但某些法律要件可以透過契約加以明確化，例如給付標的之獨特性，金錢賠償不符實際需求的情況，禁制處分亦是如此。另外，有時候不到馬上認定違約的情況，但有不安疑慮存在，也需要透過各種信用增強機制，加以保全。以下針對常見的救濟方法擇要說明。

1. 加速條款與處分即到期條款

加速條款（acceleration）用在分期給付的情況，如果一期不付，後面尚未到期的給付義務也一併到期，一次追償，不需要再等各期給付義

2　Charles M. Fox, *Working with Contracts: What Law School Doesn't Teach You* 25-28 (PLI Press, 2002). Stark, Tina L., *Drafting Contracts: How and Why Lawyers Do What They Do* 160(Wolters Kluwer, 2007).

務，到了清償期限再追償。例如：

例句八：

- -

　　If the Borrower fails to pay any Installment or Interest within 5 Business Days after the designated Payment Date, the Lender may declare the entire principal sum then unpaid immediately due.

　　處分即到期（due-on-sale）條款與加速條款有些類似，通常出現於借款擔保契約，也就是提供擔保的人如果將擔保物轉讓他人，可以宣告未清償借款全部到期。

2. 損害賠償

(1) 所受損害與所失利益

　　臺灣民法第216條第一項規定「損害賠償，除法律另有規定或契約另有訂定外，應以填補債權人所受損害及所失利益為限。」這可以作為撰寫救濟條款中損害賠償內容的第一個參考點，在臺灣，許多人撰擬契約時，也是以此作為架構損害賠償內容的基準。

　　我們可以用損害與利益這個參考座標，再加上以請求賠償的人為準、還是以被求償的人為準，將求償的內容做更細膩的區分。

　　按照請求賠償的人的觀點來看，完全履約下，他所能獲得的利益，這被稱為預期利益的損失（expectation damage），但預期利益有時候太過虛幻，仍必須有合理的限度，就像臺灣民法第216條第二項規定

「依通常情形，或依已定之計劃、設備或其他特別情事，可得預期之利益，視為所失利益。」相對的，按照賠償義務人的觀點來看，假定根本沒簽約，他就不會獲得的利益，這稱為利益返還（restitution），也就類似臺灣不當得利的概念。

按照請求賠償的人的觀點來看，根本沒有簽約的情況下，他不需要付出的成本，這稱為信賴損害（reliance damage）。相對的，假定契約完全被履行，被求償人該交付的成果，這種求償方式在美國可能透過擬制信託（constructive trust）、或設定衡平擔保權（equitable liens），也可能透過強制權利移轉（subrogation）的方式進行[註3]。不過擬制信託等方式，在被求償人看來是付出，在請求賠償的人看來則是收益。

以上是契約成立有效，但未完全履約時的處理模式。如果契約因為不符合書面要求或其他原因，未有效成立時，若一方已經有部分的履約行為，在英美法之下，就不能憑契約請求賠償，而只能依據quantum meruit的理論請求合理的給付價值。

(2) 直接損害、附隨損害與衍生損害

如果是用損害的觀點來談，則可以再區分為直接損害（direct loss）、附隨損害（incidental loss）與衍生損害（ensuing loss）[註4]，也有人稱衍生損害為consequential damage。

在英美契約法裡，直接損害的計算概念相當複雜。僅僅以買賣來說，買方的直接損害有三種計算方式，分別為cover damage、market damage，以及loss in value，而賣方的直接損害也有三種算法，包括

3　Russell Weaver, Elaine W. Shoben, & Michael B. Kelly, *Principles of Remedies Law* 122-126, 138-139 (Thomson/West 2007).

4　Russell Weaver, Elaine W. Shoben, & Michael B. Kelly 前揭著第 144-145 頁。

resale damage、market damage與loss profit。其實可以用較系統化的方式來說明這些計算方法，也就是先分為總額價差與分項價差。總額價差就是，原訂契約標的總價與另一參考價格的差額，此一參考價格可以用真的再做一次交易，也就是補買或轉賣所得到的價格（cover price, resale price）計算，也可以用其他資料建構的市場價格計算（market damage）。而分項價差則包括買方仍然收受給付標的，但請求正常標的物跟有瑕疵標的物的價值差異（loss in value），與賣方請求契約價格減除成本的利潤（loss profit）。

附隨損害指的是為了處理悔約狀況，另外支出的合理費用[註5]。以賣房子為例，賣方為了重新找買家還得多登廣告，廣告費就是附隨損害。

而衍生損害則是更比附隨損害為間接的，包括非違約方需求無法滿足所造成的損害，以及違反擔保所造成的人身財產損害[註6]。例如買方要結婚所以買房子，結果賣方悔約，買方另外找房子花了一段時間，結果還得先暫時租房子住，租金就是衍生損害。或者像是加害給付的情況，消費者買了加塑化劑的健康食品給小孩吃，愈吃愈不健康。加害給付跟侵權法有重疊的部分。衍生損害如果是交易對手難以預期的，可能無法求償。

不同的法規適用的對象有所差異，定義的附隨損害與衍生損害內容可能也有不同，例如前述附隨損害，與衍生損害的定義是美國統一商法典對貨物買賣的規範。不同的地區、或不同的交易內容很可能有完全不同的規範。在契約中，僅使用直接損害、附隨損害或衍生損害這些辭彙描述，或限制損害賠償的內容與範圍可能不夠具體。因此實務上會用例

5　U.C.C. § 2-710.
6　U.C.C. § 2-715.

示的方法進行描述，例如「including, without limitation, damages for loss of profits or revenues, business interruption, loss of business information, or other loss」，「any loss of investment, loss of contracts, loss of production, loss of profits, loss of time, loss of use or any consequential or special loss of damages」[註7]。事實上，要預先完善地設想，會出現哪些具體的附隨損害或衍生損害幾乎是不可能的任務，只能視個別交易情況，盡可能地在契約中加以描述，並加上其他限制方法，例如總賠償額度的限制。

(3) 預定損害額度的違約金與懲罰性違約金

在英美法，懲罰性違約金（penalty）的約定，在契約法上原則上是沒有效力的（但故意不實聲明，誘導他方簽約而構成詐欺，可以透過侵權行為訴訟主張懲罰性違約金）。預定損害額度的違約金（liquidated damage）（也就是作為違約時全部的賠償），如果符合一定條件，則約定有效[註8]。在臺灣，可以約定懲罰性違約金，也可以約定預定損害額度的違約金，不過違約金過高，法院可以在個案中加以酌減[註9]。中國民法典第585條亦規定了違約金的約定，但法條並未明確顯示此種違約是懲罰性的、或是預定損害額度的。較特殊的是，該條讓法院不只授權法官在個案中可以裁減違約金，也授權法官可以增加違約金。

因此，如果契約適用的法律不承認懲罰性違約金，就只能約定預定損害額度的違約金。但即使是預定損害額度的違約金也必須合情合理，

7　Fontaine & Ly 前揭著第 376 頁。

8　U.C.C. §2-718(1)規定，考量預期或實際發生的損害、損失難以證明、以及其他救濟可能不妥當等各種因素下，如果約定的金額合理，則約定有效。Restatement (Second) of Contracts §356 也有類似的建議規範。但是將損失難以證明，與其他救濟方式的不適當這兩項限制條件，局限於消費契約的情況，參閱 Russell Weaver, Elaine W. Shoben, & Michael B. Kelly 前揭著第 225 頁。

9　臺灣民法第 250、251、252 條。

不能過高，否則可能被當做懲罰性違約金，整個刪除。

　　違約金的約定方式可以是一個特定的金額，也可以是某種計算公式。

(4) 賠償金額上限

　　設定賠償金額的上限，跟預定損害額度的違約金有些相似，都是替責任額度設一個限制。不過，損害賠償金額上限只是設定上限，實際發生違約時，要賠多少，仍需要個案確定，但預定損害額度的違約金在實際發生違約事件時，通常是按照違約金所設金額求償。不論是設定損害賠償金額上限，或預定損害額度的違約金，如果個案違約情況牽涉到生命身體重大傷害，或涉及嚴重詐欺或重大過失的事件，可能責任限定的約款無法發揮效力。即使不涉及人身傷害、詐欺或重大過失，特定法域或國際條約，可能對特定交易類型的責任設有底線（例如運送人責任），契約的賠償上限約定，不能低於這些規定^{（註10）}。因此，在具體交易中應確認相關的內國法與國際法。

(5) 藍筆規則與違約金分級制

　　有的法域原則上不承認懲罰性的違約金，例如英國、蘇格蘭、愛爾蘭、美國。有的法域法官可以考量實際損失裁減過高的違約金，例如德國、法國、比利時、盧森堡、瑞士、義大利、荷蘭、葡萄牙、阿爾及利亞^{（註11）}、臺灣等。中國則是法院既可以裁減，也可以提高違約金。

　　有什麼辦法可以讓法院，不會事後更動損害賠償或違約金的約定

10　Fontaine & Ly 前揭著第382-391頁。

11　Fontaine & Ly 前揭著第342頁。

呢？其實是不容易的，因為調整違約金往往是基於法域內的公序良俗或公共政策考量。但是，在撰擬商談契約的時候，可以預先將違約金或損害賠償約定的理由，在契約說明為什麼這樣的賠償是合理的，避免爭訟時說不出理由。

此外也需要考慮個別法域究竟採用紅筆規則（red pencil doctrine）、藍筆規則（blue pencil doctrine）、或是紫筆規則（purple pencil doctrine）（註12）的考量，

在採用藍筆規則的法域，法官對於有合法性疑慮的條款只刪去有問題的部分，但不會增加或變更內容。為因應這種規則，最好將條款內容盡可能的細分。舉例來說，競業禁止的範圍避免寫一個大區，而是分成多個小區，例如約定在加州、華盛頓州、亞歷桑那州禁止競業，如果法官認為範圍過大，可以刪除其中一兩個州，但法官不會自己加上奧勒岡州，也不會自己將加州區分成北加或南加，而允許其中一部分的競業限制。又例如違約賠償條款，如果將賠償條款細分成多項，法官可刪除有疑慮的細節部分，而不是大筆一揮全部刪掉。

此外，最好把預定損害額度的違約金，按違約情事的內容與嚴重程度進行分級，如果不分違約情事輕重，一律要求同一數額的違約金，可能被法院解釋為懲罰性違約金（註13）。

(6) 談判策略

談賠償金額是極大的挑戰，交易各方商談未來合作計劃時，腦中往

12　紅筆規則指法官認為某個條款不合法，即認為整個條款無效，藍筆規則指法官認為某個條款不合法，即刪除掉不合法的部分，只刪不改，紫筆規則指法官認為某個條款不合法，可以修改增減內容。

13　Kim Lewison, *The Interpretation of Contracts* 435 (2nd ed. 1997 Sweet & Maxwell).

往浮現一片玫瑰色的未來。在這種情況下要嚴肅地談違約賠償很不容易。有時候可以換一個方向思考，不要談懲罰，而是談獎勵，也就是提前多少天履約可以提高價金（或者給提前付款的折扣）或加發獎金（註14），獎勵的成本如果事先計算清楚，那麼獎勵或處罰，只是朝三暮四、朝四暮三的修辭而已，但卻可以讓氣氛緩和。

此外，用解約款或提前終止補償款的方式來約定，或許也可避免賠償、懲罰這樣的辭彙可能造成的對立。

(7) 損害賠償條款範例

例句九：（註15）

Article 13　Breach of obligations

13.1 A Party having failed to perform properly its obligations under this Agreement may be notified by the Management Committee of this failure and invited to remedy it within a reasonable period fixed by the Management Committee {option: specify a period during which the failure must be remedied}. Any representative of the defaulting Party on the Management Committee shall not be counted as a member for this purpose. If the Party does not remedy the failure within the period fixed, it may be excluded pursuant to Article 17.

13.2 In all cases, the Party having failed to perform properly

---

14 Russell Weaver, Elaine W. Shoben, & Michael B. Kelly 前揭著第 226-227 頁。Fontaine & Ly 前揭著第 337-338 頁。

15 International Trade Centre UNCTAD/WTO, *Contractual Joint Venture Model Agreements*, 轉引自 Juris International, http://www.jurisint.org/en/con/537.html, latest accessed on 2011/6/20.

its obligations

under this Agreement shall be liable to the other Parties for the damage

resulting from its failure.

13.3 When a Party or the Joint Venture is in delay with a payment obligation,

the amount in delay shall bear interest at the average rate of actual

borrowing of the Joint Venture during the period of delay. If such average rate cannot be determined, the rate shall be the commercial rate of borrowing available to the Parties of the Joint Venture {option: provide the rate of an institution in the country of the Joint Venture, such as x points above the Central Bank's discount rate}. In case of disputes about the applicable rate, it shall be determined by the Independent Expert as provided in Article 32.8, taking into consideration the borrowing costs which the delay by a Party causes to the Joint Venture or the savings

which the Joint Venture makes by delaying the payment.

例句十：（註16）

18. Default

In default of fulfilment of contract by either party, the following provisions shall apply:

(a) the party other than the defaulter shall, at their discretion

---

16 The Grain and Feed Trade Association, *Euro-Supply Contract, General Terms*, 轉引自 Juris International, http://www.jurisint.org/en/con/65.html, latest accessed on 2011/6/20.

have the right, after giving notice by letter, telegram, telex or by other method of rapid written communication to the defaulter, to sell or purchase, as the case may be, against the defaulter, and such sale or purchase shall establish the default price.

(b) if either party be dissatisfied with such default price or if the right at (a) above is not exercised and damages cannot be mutually agreed, then the assessment of damages shall be settled by arbitration.

(c) the damages payable shall be based on the difference between the contract price and either the default price established under (a) above or upon the actual or estimated value of the goods on the date of default established under (b) above.

(d) in all cases the damages shall, in addition, include any proven additional expenses which would directly and naturally result in the ordinary course of events from the defaulter's breach of contract, but shall in no case include loss of profit on any sub-contracts made by the party defaulted against or others unless the arbitrator(s) or board of appeal, having regard to special circumstances, shall in his/their sole and absolute discretion think fit.

(e) damages, if any, shall be computed on the mean contract quantity.

例句十一：（註17）

CLAUSE 10 -UOUIDA TED DAMAGES AND BONUSES

10.1 If the Contractor fails to complete the Works within the Time for Completion in accordance with the Contract, then the Contractor shall pay to the Owner the relevant sum stated in the Appendix hereto as liquidated damages for such default and not as a penalty (which sum shall be the only monies due from the Contractor for such default) for every day which shall elapse between the Time for Completion and the date specified in the Taking-Over Certificate, subject to the limit stated in the Appendix hereto. The payment of such damages shall not relieve the Contractor from his obligation to complete the Works, or from any other of his obligations and liabilities under the Contract but shall be in full discharge of the Contractor's liability for delay in completion.

10.2 If the Contractor achieves completion of the Works prior to the Time for Completion, the Owner shall pay to the Contractor a sum as a bonus in addition to the Contract Price as stated in the Appendix hereto for every Day which shall elapse between the date specified in the Taking-Over Certificate and the Time for Completion up to the limit stated in the Appendix.

10.3 For the purpose of Clause 10 date of completion of the Works shall be the date of substantial completion thereof in

---

17 European International Contractors, *Turnkey Contract (Conditions of Contracts for Design and Construct Projects)*, 轉引自 Juris International, http://www.jurisint.org/en/con/430.html, latest accessed on 2011/6/20.

accordance with Clause 9.4.

　　10.4 If prior to the Time for Completion the Contractor in its absolute discretion following the request of the Owner agrees to allow the Owner to use or occupy the Works in whole or part then liquidated damages shall be reduced in the proportion which the value of the part so used or occupied bears to the value of the whole of the works, as applicable.

　　The provisions of this Sub-Clause shall only apply to the rate of liquidated damages and shall not affect the limit thereof,

3. 減價收受、換貨、修復與變更給付內容

　　減價收受亦是可行的救濟方式，聯合國國際貨物銷售合同公約第五十條規定「如果貨物不符合同，不論價款是否已付，買方都可以減低價格，減價按實際交付的貨物，在交貨時的價值與符合合同的貨物，在當時的價值兩者之間的比例計算。但是，如果賣方按照第37條或第48條的規定對任何不履行義務做出補救，或者買方拒絕接受賣方按照該兩條規定履行義務，則買方不得減低價格。」(註18)

　　在區域契約法及內國契約法中，將減價列為救濟方式之一也相當常見(註19)。不過，如果要在契約中將減少價金列為救濟方式，宜先談好減

---

18 第37條的救濟方式是補貨，第48條則是賣方的遲延交付獲得買方的默示諒解，可以求償，但不再減價，http://www.uncitral.org/pdf/chinese/texts/sales/cisg/V1056996-CISG-c.pdf, 最後點閱日：June 19，2011。

19 例如 Principles of European Contract Law §9:401 與 Estonian Law of Obligations Act §112, 轉引自 Peter H. Schlechtriem, *25 Years of the CISG: An International lingua franca for Drafting Uniform Laws, Legal Principles, Domestic Legislation and Transnational Contracts*, in *Drafting Contracts Under CISG* 185-186 (Harry M. Flechtner, Ronald A. Brand, & Mark S. Walter ed., Oxford, 2008)。臺灣民法第359條、中國民法典第582條也將減少價金列為救濟方式。

價的標準，這在品質與價格規格化的大宗商品可能較為容易約定。

　　換貨或修復也是可能的救濟方式，不過要考慮換貨或修復之後，如果繼續出現瑕疵該如何處理，修復或換貨應該有一個限度。除了更換相同種類的標的之外，也可以事先約定用變更給付內容方式處理。不過，變更給付內容此種方式大概較難事先規劃設計。以下提供範例：

例句十二：（註20）

8. Lack of conformity

8.5　Where the Buyer has given due notice of non-conformity to the Seller, the Buyer may at his option:

8.5.1 Require the Seller to deliver any missing quantity of the Goods,

without any additional expense to the Buyer;

8.5.2 Require the Seller to replace the Goods with conforming goods,

without any additional expense to the Buyer;

8.5.3 Require the Seller to repair the Goods, without any additional

expense to the Buyer;

8.5.4 Reduce the price in the same proportion as the value that the Goods actually delivered had at the time of the delivery bears to the value

---

20 International Trade Center, *Model Contracts For Small Firms* 49-50 (International Commercial Sale of Goods-Standard) (2011), http://www.intracen.org/model-contracts-for-small-firms/, latest accessed on 2011/6/21.

例句十三：^{（註21）}

Mandatory Price Reduction for Late Delivery. If Manufacturer fails to deliver the Products by the Scheduled Delivery Date, the Aggregate Contract Price to be paid by Purchaser is reduced by an amount equal to 1% of the original Aggregate Contract Price for each business day that the failure continues. If delivery is achieved no later than 30 calendar days after the Scheduled Delivery Date, the price reduction is the exclusive remedy of Purchaser with respect to the delay. If the delivery delay persists for longer than 30 calendar days, Purchaser is entitled to the price reduction as well as any other rights and remedies available to it under law or equity or by statute otherwise.

4. 強制履行

　　強制履行指的就是透過法院的公權力，按照原來約定的給付義務履行，例如在土地買賣契約中，賣方違約不移轉土地所有權，就訴請法院強制進行移轉。

　　臺灣的強制執行法，對於契約給付義務的強制履行並沒有太多限制。強制執行法第123條第一項規定「執行名義係命債務人交付一定之動產而不交付者，執行法院得將該動產取交債權人。」第127條第一項則規定「依執行名義，債務人應為一定行為而不為者，執行法院得以債

21　Evelyn C. Arkebauer, *Cumulative Remedies and Election of Remedies*, in *Negotiating And Drafting Contract Boilerplate* 215 (Tina L. Stark et al. ed., ALM Publishing 2003).

務人之費用，命第三人代為履行。」

　　但是在美國，並不是所有契約都能強制履行，勞務契約不管能否由第三人代為履行，法院都不會強制履行，權利人只能請求金錢的損害賠償，最多是發禁制令，禁止員工到競爭對手公司上班。而動產買賣的契約，如果是市場上就能買得到的東西，法院也不會強制履行，權利人一樣只能請求金錢損害賠償。

　　在美國，必須金錢賠償不適當，才能請求強制履行。金錢賠償是否適當有幾個判斷準則，首先是給付標的是不是罕有或獨特的，不動產大致上被認定是獨特的，而動產則必須是標的對權利人具有特殊意義、或者市場上缺貨等情況才能強制履行，否則只能請求賠錢，其次則是損害是否難以預先估計，第三是金錢求償需要連續打好幾個官司才能湊齊賠償金額、最後是交易對手財力有疑慮，可能資不抵債。因此，在個別交易中必須判斷適用的準據法，對於強制履行有無特別限制，如果有類似上述的限制，又希望未來能強制履行，最好是在契約中交代為什麼賠錢不適當，以避免對方將來抗辯賠錢就夠了。以下提供範例：

例句十四：（註22）

5.3.4 Specific Performance. Licensor acknowledges that, in the event it breaches (or attempts or threatens to breach) its obligation to provide [the services and software contemplated under this Agreement] [termination/expiration assistance] as provided in Section 6.6.3, Customer will be irreparably harmed. In such a circumstance, Customer may proceed directly to

---

22　H. Ward Classen, *A Practical Guide to Software Licensing for Licensees and Licensors* 285 (3rd ABA, 2008).

court. If a court of competent jurisdiction should find that Licensor has breached (or attempted or threatened to breach) any such obligations, Licensor agrees that without any additional findings of irreparable injury or other conditions to injunctive relief, it shall not oppose the entry of an appropriate order compelling performance by Licensor and restraining it from any further breaches (or attempted or threatened to breach).

5. 禁制處分

在涉及智慧財產授權的契約中，經常會約定一方瞭解違約可能造成的損害難以估計、或無法透過金錢有效補償，所以他方可以向法院聲請禁制處分。其用意同樣是在建立將來聲請禁制處分的條件，讓對方將來無法在法庭上做相反的主張。

例句十五：（註23）

6) Equitable Relief

(a) Any breach of this Agreement by Receiving Party will cause the Disclosing Party irreparable harm for which there shall be no adequate legal remedy. In the event of any actual or threatened breach of this Agreement by Receiving Party,

---

23 International Trade Centre UNCTAD/WTO, *Confidentiality Agreement* (2001)，轉引自 Juris International, http://www.jurisint.org/en/con/441.html, 最後點閱日：June 20，2011。

Disclosing Party shall be entitled to injunctive and all other appropriate equitable relief (including a decree of specific performance), without being required to:

(i) show any actual damage or irreparable harm,

(ii) prove the inadequacy of its legal remedies, or

(iii) post any bond or other security.

(b) If Receiving Party breaches this Agreement, Disclosing Party shall also be entitled to an accounting and repayment of all profits, compensation, and benefits directly or indirectly realized by Receiving Party as a result of such breach. Disclosing Party's remedies in this Section 6 may be exercised without prejudice to (and are cumulative with) Disclosing Party's other available rights and remedies at law, in equity, or under this Agreement, including Disclosing Party's right to monetary damages arising from any breach of this Agreement by Receiving Party.

6. 擔保品沒入、承受與換價

如果擔保品是金錢,當然可以約定有特定違約情事,即沒收擔保金。如果擔保金額合理,就不會被當作是懲罰性違約金（註24）,在英美法國家,就不會遭遇不必要的麻煩。

在臺灣,物權法修法之後,目前已經許可預先在契約中設定流質、流押條款,但需登記才能對抗第三人。在跨國交易中,如有擔保問題,

---

24 Lewison 前揭著第 437-438 頁。

仍需確認準據法是否對於流押、流質條款是否有所限制。在契約中如擬約定違約時權利人得承受擔保物，即需要預先處理擔保物如何計價以抵償違約的債務，以及相關的通知程序。至於擔保物的換價，在美國是可以透過私人自己拍賣、出租或授權以進行換價，不一定需要透過法院拍賣、或強制管理程序。同樣的，在契約中宜約定換價的方式、作業程序等事項。

在分期購物的契約中，也可能約定以銷售標的物作為擔保品，如果未依約付款，就可以沒入擔保品。此種約定一方面同樣需要考量在個別法域是否需要進行擔保交易的相關登記，如果該登記沒登記，可能不能對抗買方的其他債權人。另方面也需要考量擔保物的價值與損害額度是否相稱，例如已經繳了百分之九十的價款，最後幾期款項未繳，就把整個買賣標的所有權取回，可能失之過苛。對於此種狀況，個別法域可能有所規範。

7. 抵銷

如果雙方都互負債務、互有債權，就可以抵銷。抵銷跟同時履行抗辯的差別在於，同時履行抗辯只是擋住一時，而且要屬於同一個契約裡面的債權債務，且自己沒有先為履行的義務，才能主張同時履行抗辯。至於抵銷，依據法律，不需要屬於同一個契約，只要都屆清償期，而且給付的內容屬於同種類，就可以抵銷，抵銷的效果是消滅債權債務。但是，都屆清償期、給付內容屬於同種類是臺灣民法的限制，那是个待約定就能抵銷的情況，要特別約定才不能抵銷[註25]。在美國，普通法（也就是用判例建構的法律規則）裡沒有抵銷這回事，所以除非各州自

---

25 臺灣民法第334、335條。

已透過成文法給締約當事人抵銷權，或者當事人自己在契約裡約定，否則就沒有抵銷權了^{（註26）}。因此，在跨國交易中，抵銷權還是需要在契約中明訂。其次，法律雖然限定抵銷要都屆清償期、要給付同種類才能抵銷，但這是強行規定，還是只是沒有特別約定下的預設模式？應該只是預設模式，因為並沒有非如此不可的政策考量。所以當事人可以自己約定不同種類的給付、未屆清償期也能抵銷，只是不同種類的給付如何計價必須事先約定，否則吵到法院也難辦。以下提供範例：

例句十六：^{（註27）}

5.3.2 **Right to Set Off.** Customer shall have the right to set off any undisputed amount owed to Licensor against any damages or charges including, without limitation, Service Level Credits, assed by Customer against Licensor.

8. 利率調整

在借貸契約中也可能約定發生違約事件，就調漲借款利率，如果約定的利率調漲幅度過大，或者約定可以溯及既往地調高利率，可能在英美法國家被解釋為懲罰性違約金，從而無效^{（註28）}。

26　Classen 前揭著第 67 頁。
27　Classen 前揭著第 279 頁。
28　Lewison 前揭著第 438-439 頁。

9. 不安抗辯與信用增強

有時候如果等到債務人跳票，才採取行動，可能已經太晚。因此必須更早掌握債務人信用能力的變化情況，一旦有重要變化，就要採取行動。但是把這種重要變化，列為違約事件或契約終止解除事由可能太過急進，可以把這樣的事件列為讓債權人要求增加擔保或行使不安抗辯的理由。

臺灣民法第265條規定「當事人之一方，應向他方先為給付者，如他方之財產，於訂約後顯形減少，有難為對待給付之虞時，如他方未為對待給付或提出擔保前，得拒絕自己之給付。」因此，可以用對方的財產顯形減少，作為要求增加擔保或我方暫停履約的事由。

美國統一商法典第2-609(1)條規定，一方在訂約後發生常理看來不太安全的情況（insecurity），他方得以書面要求增加擔保。

在契約裡，我們當然不能只把法條文字抄上去，而是要定義哪些事是「有難為對待給付之虞」的事，哪些事是不安全的狀況。否則，等發生事情再到法院吵，不一定吵得贏。此外，擔保的額度、擔保的種類如何搭配重大信用能力變化，如果能列出級距表，那是更完善了。

要瞭解交易對手的重大信用能力變化，除了看報紙、看新聞以外，更要在主要給付事項條款裡面，設定交易對手接受詢問、回答問題、接受定期、不定期稽核、檢查的義務，也要設定交易對手定期主動提供資料的義務。有些法人貸款契約，甚至還會要求讓銀行派人到借款公司擔任董事，直接在內部掌握第一手資訊。當然，各種管控手段也不能太超過，變得礙手礙腳。

第五節　救濟條款其他相關事項

　　除了列出違約事件、救濟內容以外，救濟條款也可以約定舉證責任的分配。有時候者種舉證責任的分配，是透過較間接的方式進行的，例如在法國，如果契約約定必須達成特定結果的給付義務（obligations to achieve specific result），那麼義務人要證明他的給付內容符合約定，但如果是勤勉努力義務（obligations of due diligence）就變成權利人要舉證義務人的作為，不符合勤勉努力義務的要求[註29]。不過還是要留意，太間接曖昧的效果，不只交易對手可能不瞭解，變成只是保留吵架本，而不是真的有益於雙方誠信履約，吵到法院時，法官也不一定買帳。

　　另外，救濟條款時常也會限制求償期限或做其他形式要求。這些限制當然不是任何一方可以為所欲為，還是要合情合理。

　　救濟條款也必須說明，各項救濟內容是只能選擇一項主張，還是互不排斥，可以同時請求。如果希望限制救濟內容，要求權利人只能擇一主張，需特別留意契約中明文限制的對象是賠償（damage）？還是救濟（remedy）？如果限制的對象是賠償，那只有金錢賠償的部分被限制了，權利人除了要求賠償金，可能還可以依據契約或法律規定主張其他救濟，例如indemnity、終止契約、沒收擔保金。如果希望限制救濟選項，就要寫成「[Remedy Content] is the exclusive remedy of [Party Name]'s for [Specific Default]」。[註30]

　　此外，特定法域的法律，對於救濟內容極可能也有相關規定，那麼法定的救濟內容跟契約約定的救濟內容是能夠併存的，或是必須選擇其

29　Fontaine & Ly 前揭著第 366-367 頁。
30　參閱 Arkebauer 著作第 218 頁。

一行使權利，或者是完原排除法定救濟內容，只能按照契約救濟條款行使權利，必須寫的非常清楚。按照美國統一商法典的規定，如果不寫清楚，就當作契約約定的救濟內容只是多給的，並不影響當事人依據法律規定請求救濟的權利（註31）。

---

31 U.C.C. §2-719(1)(b). Russell Weaver, Elaine W. Shoben, & Michael B. Kelly 前揭著第224頁。

散場條款

天下無不散的宴席。我們之前提過契約最重要的四個內容：來者何人、有何指教、珍重再見，以及在哪裡輸贏。所以，契約的轉折也需要有散場條款（Endgame Provisions）。而且散場條款非常重要，交易如果變調虧錢還不能走人，就像怨偶勉強在一起，可能毀掉一生，而分手分得不好，也會留下很多麻煩事。

第一節　撤銷、撤回、終止、解除等不同概念的區別

意思表示或契約關係的結束，涉及多個相近的概念：撤銷（rescind）、解除（repudiate）、終止（terminate）、撤回（revoke）。英文與中文的對應文字只能說是較常見的，但不是完全不變的，因為使用英文的國家很多，有的國家使用英語就是不同於多數情況。我們用表格來解釋這四個概念。

表1　撤銷、撤回、終止、解除概念對照表

	意思表示	契約
自始無效	撤銷、撤回	撤銷、解除
向後失效		終止

意思表示就是一方想要締結某個契約的意思，但是契約原則上要兩個意思表示相互合致才能成立契約。例外情況是一方的意思表示加上另一方實際履約的行動，這稱為諾成契約（unilateral contract），例如演藝公司辦試鏡選角活動，廣發消息，意者自錄影片，報名參加，優勝的才有獎。所以要得到獎賞，對價就是提供影片，現場演出，而且被評選最佳，這在臺灣稱為優等懸賞廣告。

我們回到締約的原則方法，契約是由雙方當事人兩個意思表示構

成，兩個意思表示原則上一前一後，而且內容一致，第一個意思表示稱為要約，第二個意思表示稱為承諾。

撤銷可適用在意思表示、也可適用在契約，其效果是讓意思表示或契約自始無效。而之所以可以撤銷，並不是契約約定，而是法律規定的，通常的原因是意思表示有問題，例如未成年人行為能力不健全、例如意思不清醒、不自主，像是喝醉酒、嗑了藥、被詐欺、或被脅迫。如果還沒成立契約，就撤銷意思表示，如果已成立契約，就只能撤銷契約。撤銷通常是撤銷自己的意思表示或契約，但例外情況也可以聲請法院撤銷別人的契約，例如債務人為了脫產，而跟第三方做假交易，賤價或無償轉讓原本債權人可能執行的財產。

解除的效果是自始無效，對象則限於契約，可能是契約約定有解除權，可能法律規定有解除權，通常是一方違約，或有破產等財信惡化之狀況。

終止的對象也限於契約，效果是向後失效，就像租房子，提早結束租約。只有繼續性的契約才能終止。可能契約中約定終止權，或者法律規定有終止權。

撤回則只針對意思表示，契約成立後就不能撤回意思表示了。撤回不需要原因，只要夠快就好。在大陸法系，要快到原來的意思表示還沒到，一言既出，駟馬難追，要約發出之後，如果對方已收到要約，就要給對方一段合理期間考慮參詳，過了合理期間仍未承諾，要約就失效了，也無須撤回。但如果發出要約之後，立即多派幾匹馬追得到，就可以撤回。Line對話，在對方已讀之前趕緊收回，就是撤回。在英美法系，比較不講武德，要約可以任意撤回，就算對方已收到意思表示，只要還未承諾，都可撤回，所以下了訂單，只要還未承諾，都可以隨時抽回要約。

　　契約的散場條款通常包括契約終止或解除 (註1)。撤銷或撤回通常
都是基於強行適用的法律規定，契約沒辦法變更強行適用的法律，所以
也不用在契約裡約定撤銷或撤回。

　　終止或解除也可以只針對契約的部分內容，而是全部終止或解除，
完全由當事人自行視情況需要協議。以下說明終止或解除條款的寫法。

第二節　契約終止或解除的關鍵動詞

　　契約終止指的是持續履行的契約提前結束。例如本來租房子要租一
年，發生特殊情況，所以租了半年就終止租約，但前半年交了房租，住
了房子都是有效的。在英文契約中，終止可以用「terminate」來表示。

　　終止跟解除是不同的，解除契約指的是契約根本就當作從來沒發生
效力。例如本來約定六月一日房東要親自交屋，結果六月一日舊的房客
還不搬走，沒辦法交屋，所以新的房客不租了，解除契約。在英文契約
中，解除可以用「rescind」來表示，避免使用「cancel」，因為美國統
一商法典第2-106條將「termination」定義為非違約事件的情況下停止
契約效力，而且是向後失效。「cancellation」則定義為違約情況下停止
契約效力，效果跟「termination」一樣，只是多了賠償問題。但
「rescind」則有讓契約打從開始就無效的意思，也才跟臺灣法下解除的
概念一致 (註2)。

1　Tina L. Stark, *Drafting Contracts: How and Why Lawyers Do What They Do* 157-163 (Wolters Kluwer, 2007). 紛爭解決條款可以放在散場條款中處理，也可以放在一般條款處理，本書將在一般條款中再行討論紛爭解決條款，因為糾紛的來源不僅有違約事件。

2　在臺灣，對於撤銷與解除加以區別，但在美國，對這兩種概念並無明確的區分，都可以用rescind表示。

第三節　契約終止或解除的原因

　　契約終止或解除通常是用裁量選擇句表示，也就是用「[Party Name] may terminate/rescind this Agreement…」表示在一定情況下，締約當事人一方或雙方，有裁量權可以用某種方式來終止或解除契約。盡量不要用宣示句寫成自動終止或解除，因為那樣可能太僵硬而沒有彈性，有時即使發生不好的狀況，仍想維持契約的存在。

　　終止的權限可能只給當事人一方，也可能雙方都有權終止。但解除通常是較為嚴重的情況，所以較少約定雙方當事人都能解除契約。此外，也可以設定支付代價而終止或解除契約，這種約定某程度上取代了違約金[註3]。

　　契約終止或解除的原因有時候是對方違約，有時候是不能歸咎任何一方的情況，有時候反而是自己發生某種特殊情況，想要提前終止或解除。因此，按照終止或解除的原因，可以大致區分成兩類，一種是任意（「at will」或「without cause」），另一種是附理由的終止（with cause）。任意即不再過問理由。如果事先許可任意終止或解除，就只要求一定時間前用特定方式通知，或者加上一些補償。至於附理由的終止或解除，可能是一方的違約、資不抵債、破產、重整、不可抗力造成無法繼續履行、履約目的受挫或某種僵局（例如合資設立事業，之後產生無法調解的意見紛歧）、或是契約主要目的提前達成。[註4]

　　此處要留意，不要替自己設計可以任意終止或任意解除，而且一通電話，立刻生效，也不用任何補償的終止或解除條款。因為在英美契約

3　Marcel Fontaine & Fillip De Ly, *Drafting International Contracts: An Analysis of Contract Clauses* 336-337 (Martinus Nijhoff 2009).

4　Fontaine & Ly 前揭著第 569-570 頁。

法之下，這麼做可能會產生約因的問題。契約裡約定的各項義務，如果一通電話就全部煙消雲散，等於自己沒有拿出對價跟對方交換。沒有約因，就不成立契約，雙方都可以不玩。所以，最好加上個幾日前通知，有個合理的時間限制，才不會發生問題。由此也可以瞭解，契約還是要誠信，不能機關算盡，欺人太甚。各種法律原則都是要確保當事人在誠信的原則下為自己的利益做安排。

至於附理由的終止或解除，可以跟各項給付義務放在一起，也可以集中放在散場條款裡的契約終止。不過並不是所有的終止事由都來自於違約，例如以下例句一2.2.1提到被政府主管機關命令停止執行某項研究計劃。如果只在給付義務或聲明擔保的相關條款中附帶說明，很可能掛一漏萬。因此，集中成終止事項條款較為明確而妥當。

第四節　契約終止或解除後的善後

終止或解除之後可能有些善後問題需要在終止事項中直接處理。例如高階經理人被任意解職，通常有巨額的離職金，這稱為「黃金降落傘」（golden parachute）。而某些善後問題可能在其他個別條款中交代（例如：租屋終止後回復原狀、繳清水電、瓦斯費）。

任意終止或解除跟附理由的終止或解除有時候會併存，但是有不同的效果。例如在聘僱高階經理人的情況，如果高階經理人不盡忠職守，自肥營私，當然公司可以開除，不但不給離職金，甚至還要追究賠償責任。但有時候高階經理人已經盡力經營，水土不服、時不我予，還是該給離職金，好聚好散。

另外需留意有時候法律會特別規範契約終止後的補償，例如臺灣的

勞基法，資遣員工要給資遣費。在歐洲多個國家，也有規定結束代理合約，不論是到期結束，或者提前終止，在符合某些要件時，必須付給代理商一筆補償費（註5）。所謂代理本來指的是代理商去招攬顧客，向委任的廠商報告這樣的締約機會，成交之後委任的廠商付給代理商佣金這樣的交易型態。但一些歐洲國家將這種補償代理商的法律規定也類推適用到經銷商，所謂經銷商，是指向供應商購買產品，經銷商再轉售給終端客戶的交易型態。因此，需要特別留意以上規定，事先透過管轄地、準據法的約定以及交易關係的設計，以有效規避這種補償規範。

第五節　限制終止或解除權的法律規定

也須留意，有些終止條款可能沒有法律效力。例如在美國，依據破產法（註6）規定，待履行契約或者未到期的租約即使約定當事人一方破產，他方可以終止契約，這樣的約定是沒有效力的。所以要更準確的掌握交易對象的財務狀況，及早因應。

以下提供例句。

5　例如瑞士債法第418u條

1 Where the agent's activities have resulted in a substantial expansion of the principal's clientele and considerable benefits accrue even after the end of the agency relationship to the principal or his legal successor from his business relations with clients acquired by the agent, the agent or his heirs have an inalienable claim for adequate compensation, provided this is not inequitable.

2 The amount of such claim must not exceed the agent's net annual earnings from the agency relationship calculated as the average for the last five years or, where shorter, the average over the entire duration of the contract.

3 No claim exists where the agency relationship has been dissolved for a reason attributable to the agent.

https://www.droit-bilingue.ch/en-it/2/22/220-index.html, latest accessed on September 1, 2023.

6　11 U.S.C. 365 (e)(1)，轉引前註1Stark 著作第161頁。

例句一：（註7）

2.2. – This Agreement may be terminated prior to completion of the Study as follows:

2.2.1. – CHEMICAL or INSTITUTION may terminate this Agreement immediately if required to halt the Study by FDA or compelled to do so by reasons of subject safety;

2.2.2. – CHEMICAL or INSTITUTION may terminate this Agreement at any time upon thirty (30) days, written notice;

2.2.3. – CHEMICAL or INSTITUTION may terminate this Agreement upon thirty (30) days' written notice in the event of a breach by the other party of its obligations under this Agreement and a failure by the other party to correct its breach(es) within that thirty (30) day notice period; or

2.2.4. – CHEMICAL or INSTITUTION may terminate this Agreement in the event that Principal Investigator can no longer serve as principal investigator and the Parties cannot agree upon an available, acceptable substitute.

2.3. – In the event of early termination of this Agreement, INSTITUTION shall immediately return any unspent and uncommitted portion of any payments made by CHEMICAL pursuant to Section 1.4. In the event of early termination of this Agreement by either party pursuant to Sub-Sections 2.2.2, or by 2.2.4, by CHEMICAL pursuant to Sub-Section 2.2.2, or by INSTITUTION pursuant to Sub-Section 2.2.3, CHEMICAL shall

---

reimburse INSTITUTION for all reasonable expenditures on materials supported by invoices that were incurred prior to notice of termination, subject to the maximum amount to be paid under Section 1.4 above.

例句二：

XII. Early Termination

In the event that either party believes that the other materially has breached any obligations under this Agreement, or if Licensor believes that Licensee has exceeded the scope of the License, such party shall so notify the breaching party in writing. The breaching party shall have [time period] from the receipt of notice to cure the alleged breach and to notify the non-breaching party in writing that cure has been effected. If the breach is not cured within the [time period], the non-breaching party shall have the right to terminate the Agreement without further notice.

Upon Termination of this Agreement for cause online access to the Licensed Materials by Licensee and Authorized Users shall be terminated. Authorized copies of Licensed Materials may be retained by Licensee or Authorized Users and used subject to the terms of this Agreement.

In the event of early termination permitted by this

8　Yale University, *Standard License Agreement*, 2001,轉引自 Juris International, http://www.jurisint.org/en/con/433.html, latest accessed on 2011/6/11.

Agreement, Licensee shall be entitled to a refund of any fees or pro-rata portion thereof paid by Licensee for any remaining period of the Agreement from the date of termination.

例句三：（註9）

12. Consequences of Termination

12.1 Upon the termination of this Agreement for any reason:

12.1.1 The Manufacturer shall be entitled (but not obliged) to repurchase from the Distributor all or part of any stocks of the Products then held by the Distributor at their Invoice Value or the value at which they stand in the books of the Distributor, whichever is lower; provided that:

a) the Manufacturer shall be responsible for arranging and for the cost of, transport and insurance; and

b) the Distributor may sell stocks for which it has accepted orders from customers prior to the date of termination, or in respect of which the Manufacturer does not, by written notice given to the Distributor within 7 days after the date of termination exercise its right to repurchase, and for those purposes and to the extent the provisions of this Agreement shall continue in full force and effect;

12.1.2 The Distributor shall at its own expense within 30 days send to the Manufacturer or otherwise dispose of in

---

9   Chamber of Commerce and Industry of Western Australia, Distributorship Agreement，轉引自 Juris International, http://www.jurisint.org/en/con/385.html, latest accessed on 2011/6/20.

accordance with the directions of the Manufacturer all samples of the Products and any advertising, promotional or sales material relating to the Products then in the possession of the Distributor;

12.1.3 Outstanding unpaid invoices rendered by the Manufacturer in respect of the Products shall become immediately payable by the Distributor and invoices in respect of Products ordered prior to termination but for which an The Distributor shall at its own expense within 30 days send to the invoice has not been submitted shall be payable immediately upon submission of the invoice;

12.1.4 The Distributor shall cease to promote, market or advertise the Products or to make any use of the Trade Marks other than for the purpose of selling stock in respect of which the Manufacturer does not exercise its right or repurchase;

12.1.5 The Distributor shall at its own expense join with the Manufacturer in procuring the cancellation of any registered user agreements entered into pursuant to Clause 7.6;

12.1.6 The provisions of Clauses 8 and 9 shall continue in force in accordance with their respective terms;

12.1.7 The Distributor shall have no claim against the Manufacturer for compensation for loss of distribution rights, loss of goodwill or any similar loss; and

12.1.8 Subject as otherwise provided herein and to any rights or obligations which have accrued prior to termination, neither party shall have any further obligation to the other under the Agreement.

紛爭解決條款

如同之前說明的，契約最重要的四個內容：來者何人、有何指教、珍重再見，以及在哪裡輸贏。契約中的紛爭解決條款就是要處理發生糾紛時如何決勝負。如果紛爭解決條款設計不當，常常是未戰先敗。例如臺灣公司與英國經銷商簽訂契約，約定糾紛專屬英國法院管轄，之後雙方發生糾紛，臺灣公司表示英國法院歧視有色人種而不公，而在英國開庭程序中擬聘用翻譯也受阻撓。(註1)紛爭在何處處理至關重要，就像江湖傳聞六大門派圍攻光明頂為什麼會輸，因為光明頂在新疆的山上，六大門派要跋山涉水過去太遠了。

紛爭解決條款的內容通常包括契約準據法的約定與紛爭處理機制的約定。以下分別說明。

第一節　契約的準據法（Governing Laws）

不同法域的當事人締結的契約如果未約定準據法，發生糾紛時，適用哪一個法域的法律可能會因為在哪個法庭起訴而適用不同的法律衝突法則（principles of conflict of laws），最終適用不同法域的法律。就像新聞常常報導，不同國籍的藝人結婚，未簽訂婚前協議，鬧離婚時，也談不攏在哪裡訴訟解決，可能會因為哪一方先到哪一個法院起訴，而影響婚姻的準據法。

所以不同法域的當事人締約，最好約定清楚以哪一個法域的法律為準據法。

當事人時常因為爭執究竟該適用哪一個法域的法律，僵持不下，無法達成共識，而互相妥協選擇第三地的法律，但卻根本不了解第三地法

1　一堂十億元的教訓 必翔纏訟十年，慘賠十億的孤兒心聲, https://news.pchome.com.tw/magazine/report/po/new7/7184/133104960060429001001_2.htm, latest accessed on August 31, 2023.

律，從而造成選定的準據法無法適用。

　　假設臺灣公司跟日本公司做商品買賣交易，交易金額一百萬新台幣，契約明定適用美國紐約州法，後來才發現紐約州法規定，如果當事人間的這項交易跟紐約州基本上沒什麼關係，不能隨便選用紐約州的法律作為準據法，必須交易金額達到二十五萬美金以上，才能選用紐約州法律作為準據法（註2），當事人一方日後在法庭上提出此一事實，到底該如何解決？

　　臺灣涉外民事法律適用法第二十條第一項規定：「法律行為發生債之關係者，其成立及效力，依當事人意思定其應適用之法律。」，並未限制契約內容必須跟當事人選定的準據法地有何牽連關係。但是某些法域可能要求當事人的契約與選用的準據法地有實質關係或必須具備其他合理的基礎，且不違反真正具實質利害關係的法域的公共政策（註3）。

　　除了上述一般的效力問題之外，個別實體法律對於特定交易行的準據法亦可能有強制規定，無法另行約定準據法。例如在美國要進行擔保交易，如何完成對抗第三人要件（perfection, 例如：登記），要依據債務人住所地的法律判斷。美國統一商法典對於商品出賣人的無擔保債權人，對於出賣人已經選定要交付給買受人的商品，能否主張買賣行為是詐害行為，而屬無效，要依據商品所在地的法律（註4）。

　　因此，在具體選定準據法之前必須依照個別的交易內容，審慎考量選定的準據法究竟能不能用？就算選定準據法有效，也需要考慮跟管轄法院的選定能不能搭配，如果約定在臺灣打官司，卻用美國法律，如何

2　N. Y. General Obligations Law §5-1401(1).

3　*Restatement (Second) of Conflict of Laws* § 187(2) (1971).

4　Brad S. Karp & Shelly L. Friedland, *Governing Law And Forum Selection*, in *Negotiating And Drafting Contract Boilerplate* 117 (Tina L. Stark et al. ed., ALM Publishing 2003).

將美國法律調查清楚，然後向臺灣法院完整說明，也是一件難度甚高的工作。

如果準據法選定條款是有效的，接著就必須考慮選定某一地區的法律，對於此項交易日後可能產生的糾紛有何影響？可能某個地區的法律對於某一種類型的當事人較為嚴苛，例如在紐約州，投資公司以外的未上市、櫃的股份有限公司前十大股東對於員工的薪資要負連帶責任〔註5〕。這些玄機都必須按照交易內容具體確認，否則選定一個對自己非常不利的法律就很麻煩了。

即使約定了準據法，在糾紛發生時，對造仍可能主張適用其他法律。例如臺商在大陸設公司，與美國的公司進行貨物買賣，由於大陸與美國都是聯合國國際貨物銷售合同公約的簽約國，因此依據公約第一條就適用公約的規定。當然，雙方可以依據公約第六條特別約定排除公約的適用。不過，如果締約當事人在契約中只寫適用特定地區的法律，是否算是默示排除公約的適用呢？可能在個案訴訟中引發糾紛〔註6〕。因此在約定準據法時，宜明確排除選定準據法與法院地程序法以外的其他地區法律，或跨國公約規範。

選定準據法時也必須留意是否應排除衝突法，以避免造成反致的問題。舉例而言，如果在契約中約定適用加州法，結果依據加州的衝突法規定，本項交易糾紛應適用其他地區的法律，這就是一種反致。因此，在契約準據法條款中，最好是排除選定法域衝突法的適用，以避免複雜化。

以下提供準據法條款的範例：

5　N.Y. Business Corporation Law §630 (a).

6　Michael Bridge, *Choice Of Law And The CISG: Opting In and Opting Out* in *Drafting Contracts Under The CISG* 76 (Harry M. Flechtner et al. ed., Oxford 2008).

例句一： (註7)

Governing Laws. The substantive laws of the [Country Name] (without giving effect to its conflicts of law principles) govern all matters arising out of or relating to this Agreement and all of the transaction it contemplates, including, without limitation, its validity, interpretation, construction, performance, and enforcement. Except for the substantive laws of [Country Name] and the procedural laws of forum agreed by the Parties, all other state, national, transnational laws conventions, and treaties are excluded from application.

第二節　紛爭處理機制

紛爭處理機制種類繁多，可以當事人自己協商（negotiation），可以找第三方調解（mediation），可以循專家裁決（expert determination）。但是最普遍採用的仍然是訴訟（litigation）與仲裁（arbitration）。協商與調解不一定有結論，都是要當事人自願接受協商調解結果，才會有拘束力。而專家裁決適用在專業領域，有該領域專家的情況。訴訟與仲裁必然會有明確且拘束當事人的結果，也不局限必須找到專家。

訴訟就是上法院請法官做判決，仲裁是由爭議當事人自行指定仲裁人，自行決定仲裁程序（ad hoc arbitration），或委託仲裁機構按機構的程序來仲裁（institution arbitration），就爭議做出有拘束力的仲裁判

7　參閱 Brad S. Karp & Shelly L. Friedland 前揭文第 120 頁。

斷。法院是政府機構，法院的程序較固定難有彈性，法官領政府固定薪
水，日理萬機，案牘勞形，對商務運作可能比較不熟稔，而且法院可能
有多個審級，可能經歷數十年還沒辦法解決。仲裁人或仲裁機構都是個
案受託並支領報酬，程序較有彈性，而且當事人可以找個別商務領域的
公正人士擔任仲裁人，信賴關係與互動關係通常會比法官與當事人好，
仲裁通常是一級程序就解決了，而且仲裁庭通常六個月到一年之間一定
會做出判斷，對仲裁判斷只有在很特殊的情況才能表示不服。法院的審
判程序原則上是公開的，裁判書也會上網，而仲裁程序與判斷是不公開
的。但法院訴訟的優點是判決確定後即可聲請強制執行，仲裁判斷則須
經由法院認可才能聲請強制執行，認可程序又需要耗費時間。務須留
意，在契約中，訴訟或仲裁只能擇一寫上，不能寫「可訴訟或仲裁」，
否則一方提訴訟，另一方提仲裁，到底哪一個程序有效，就有得吵了。
以下我們分別說明契約中的管轄法院條款、仲裁約定機制條款、與其他
解決機制。

1. 管轄法院（Jurisdiction or Venue）

　　管轄法院的約定受到各法域法律與公共政策的控制，未必能夠任意
選擇。另外如果選定的法院對當事人一方，甚至對法院來說在調查證
據、傳喚證人上也極為不便，則管轄法院的約定可能無法發揮效力。管
轄法院可能以「不便利法庭」（forum non conveniens）的理由不受理
案件，因此需要預先考慮個別契約與管轄法院所在地區有何牽連關係，
瞭解該地的衝突法與訴訟法相關規定。一造在約定的管轄法院起訴，他
造還是可以抗辯管轄法院不適當，並非約定了，就有絕對的拘束力。所
以需要特別寫明當事人放棄日後抗辯管轄不適當的權利，個別法院對於
放棄抗辯權聲明的內容格式可能有所規定，所以要找出個別法院的書類

範例。（註8）

　　紐約州對於本來不能在該州提出的契約糾紛訴訟，也定出一百萬美金的門檻，交易金額達到這一數額，就可以約定由紐約州法院管轄。（註9）所以也需留意此類的管轄門檻。

　　選擇管轄法院除了考量語言、區位等應訴便利性因素，以及法院公正性之外，也需要考量程序的複雜程度與費用。美國民事訴訟程序跟臺灣的民事訴訟程序比較之下有三大特點。如果約定在美國打官司，就必須因應這些特點，約定清楚。

　　首先是事證開示（Discovery）制度，就是雙方自己先約定時間，調取他方文件檔案，或詢問他方相關人員作成筆錄（庭外取證，deposition），通常是由法院書記在場主持。一開始，事實關係未必明確，如果由法官直接指揮釐清事證的程序，可能很沒效率。事證開示制度有助於釐清事實、證據與爭議事項，提升裁判品質。但相對的，雙方在調查證據或配合對方調查證據上，就要付出很多的時間與金錢成本。

　　其次，美國的民事案件在一定條件下也可以選擇交由陪審團審理，締約雙方對於未來糾紛是否要交由陪審團審理，也需要預先約定清楚。對臺灣公司來說，最好是不要在美國跟美國公司玩陪審團。一方面人不親土親，陪審員可能偏好美國公司，再者，美國公司對陪審團運作比臺灣公司熟稔太多，而且陪審團的費用與時間成本都很高。美國有些州不允許在訴訟發生之前就約定放棄找陪審團審案權利的，因此跟那些州的公司做生意，如果絕對不想要找陪審團，就必須在契約中排除那些州的管轄（包括聯邦法院在該州的地區法院）的管轄權。其他各州均許可預先聲明放棄陪審團審案權利，但是聲明放棄的意思必須表達得非常明確

8　改編 Brad S. Karp & Shelly L. Friedland 前揭文第 135-137 頁示範例。
9　N. Y. General Obligation Law § 5-1402.

是出於完整認知，且自願的。

最後是送達，美國訴訟書狀的傳統送達模式是當事人自己找律師或專門送達的服務機構親自送給對造，書狀沒送到可能審判程序就玩不下去了，不是隨便找郵局寄過去就可以了。所以，如果約定在美國打官司，要一併約定清楚豁免掉這種傳統模式的送達，可以用郵局掛號信或其他方式送到當事人自己指定的地點與收件人。

管轄法院的選擇有兩種模式。第一種是專屬的合意管轄，也就是發生糾紛時，只能向約定的法院起訴。第二種是非專屬的合意管轄，也就是除了選定的法院以外，還可以依據其他法律規定，找適當的法院提起訴訟。如果是非專屬的合意管轄，事後可能就要比賽誰先提起訴訟，先提起訴訟的一方可能就決定了在哪裡打官司。以下提供管轄法院約定條款的範例。

例句二：（註10）

Forum. [Court's Name] has exclusive jurisdiction for all legal action or proceeding arising out of or relating to this Agreement or the transactions it contemplates. Each party waives to the extent permitted by law, (a) any objection to the laying of venue of any legal action or proceeding arising out of or relating to this Agreement brought in above agreed court, and (b) any claim that above agreed court is an inconvenient forum.

Process Agent.

(i) [One Party] irrevocably appoints [One Agent's Name] as

---

10 改編 Brad S. Karp & Shelly L. Friedland 前揭文第 136, 140, 142 頁。Lauren Reiter Brody et al., *Waiver of Jury Trial* in *Negotiating And Drafting Contract Boilerplate* 163 (Tina L. Stark et al. ed., ALM Publishing 2003).

its agent to receive service. Any process service from [The Other Party] to [One Party] is improper, except for those made by certified mail to [One Agent's Name] at [One Agent's Address], or any other address [One Agent's Name] notifies [The Other Party].

(ii) [The Other Party] irrevocably appoints [The Other Agent's Name] as its agent to receive service. Any process service from [One Party] to [The Other Party] is improper, except for those made by certified mail to [The Other Agent's Name] at [The Other Agent's Address] or any other address [The Other Agent's Name] notifies [One Party].

(iii) Nothing set forth in this Section affects the right to serve process in any other manner permitted by law.

Waiver of Jury. Each party, to the extent permitted by law, knowingly,

voluntarily and intentionally waives its right to a trial by jury in any action or other legal proceeding arising out of or relating to this Agreement and the transactions it contemplates. This waiver applies to any action or legal proceeding, whether in contract, tort or otherwise.   [One Party]_____(initial) [The Other Party]_____(initial)

2. 仲裁（Arbitration）

相較於訴訟程序，仲裁程序有許多優點，例如成本較低、速度較快，也較能維持隱密，而且可以挑選有相關專業背景的仲裁人。但相對

的，仲裁判斷可能不是嚴格依據法律規定，較難預測結果。

　　所以，締約雙方應該預先考慮周全究竟要不要採用仲裁。當然，也可以區分某些糾紛適用仲裁，某些糾紛適用訴訟，但這種做法執行上有相當多的困難。一方面，未發生糾紛之前要預先想到哪裡會出現糾紛是不容易的，二方面，多個糾紛議題之間要劃出清楚的界線也很難，雙方很可能為了特定事項屬不屬於約定仲裁事項而產生爭執，三者，一旦同時發生多項糾紛，要同時跑法院與仲裁機構，個別進行不同的程序也會造成困擾。

　　決定仲裁與否，以及仲裁事項之後，締約當事人也必須要選擇一個仲裁地點、仲裁機構，並考慮仲裁程序該由仲裁機構統籌管理，還是締約當事人自主安排的仲裁程序。如果交由仲裁機構統籌管理仲裁程序的進行，包括雙方如何交書面資料，仲裁審議的時間、地點，進行程序等等，自然是較方便。締約當事人自主安排的話，很可能在程序進行上卡住了，最後還是必須到法院解決。不過即使交由仲裁機構統籌管理，當事人還是可以針對特定事項約定，例如證據調查的方法，能傳喚多少證人，是否進行庭外取證，都可以客製化。

　　接著就必須考量選任仲裁人的問題，如果當事人不約定仲裁人選任方法，就會按仲裁機構的規則處理。仲裁人的專業資格與背景也可以加以限定。仲裁人的人數通常是指定一人或三人，簡單案件一個人，複雜案件三個人，原則上可由兩造各挑一個，當事人挑出來的兩位仲裁人再自己協商決定第三位仲裁人。

　　仲裁人最終判斷的權限也可能需要限定，例如能不能做懲罰性違約金的判斷，能不能做暫時處分，利息怎麼決定，判斷的賠償金額是否設上限，提起仲裁有無時效限制，仲裁判斷是否須附理由，以及仲裁費用如何負擔等。暫時處分可以約定向法院聲請，如果仲裁機構的規則許可

仲裁庭做出暫時處分，也可以約定向仲裁庭聲請。向法院聲請是較保險的做法，法院一定有此權限，而且在暫時處分的執行上也較便利。

最後則是仲裁判斷的執行與上訴問題。^{（註11）}

以下提供仲裁條款範例，在撰擬個案仲裁條款，可參閱選定仲裁機構的網站提供的仲裁條款範例，加以調整。

例句三：^{（註12）}

Article X

Arbitration

Section X.01. Scope of Arbitration. Any controversy or claim arising out of or relating to this Agreement is to be resolved by arbitration.

Section X.02. Administration of Arbitration. The arbitration is to be administered by the American Arbitration Association and is to be conducted in accordance with the Commercial Arbitration Rules of the American Arbitration.

Section X.03. Appointment of Arbitrators. The Arbitration is to be held before a panel of three arbitrators, each of whom must be independent of parties. No later than 15 days after the arbitration begins, each party shall select an arbitrator and request the two selected arbitrators to select a third neutral arbitrator. If the two arbitrators fail to select a third on or before

---

11　Elliot E. Polebaum, *Arbitration*, in *Negotiating And Drafting Contract Boilerplate* 181-199 (Tina L. Stark et al. ed., ALM Publishing 2003). Larry A. DiMatteo, *The Law Of International Contracting* 46-51 (Kluwer Law International 2000).

12　改編 Polebaum 前揭文第 193，201-202 頁與 Larry A. DiMatteo 前揭著第 51 頁示範例。

the 10th day after the second arbitrator is selected, either party is entitled to request the American Arbitration Association to appoint the third neutral arbitrator in accordance with its rules. Before beginning the hearings, each arbitrator must provide an oath or undertaking of impartiality.

Section X.04. Scope of Arbitrators' Authority.

(a) *Interim Relief.* Either party may seek from any court having jurisdiction any interim or provisional relief that is necessary to protect the rights or property of that party. By doing so, that party does not waive any rights or remedy under this Agreement. The interim or provisional relief is to remain in effect until [an arbitral tribunal is established] [the arbitration award is rendered or the controversy is resolved].

(b) *Punitive Damages.* The arbitrators have no authority to award punitive damages or other damages not measured by the prevailing party's actual damages.

Section X.05. Escrow. Pending the outcome of the arbitration, [name of party] shall place in escrow with [law firm, or arbitration institution], as the escrow agent, [the sum of ___, a letter of credit, goods, or the subject matter in the dispute]. The escrow agent is entitled to release the [funds, letter of credit, goods, or subject matter in dispute] as directed by the arbitrator in the award, unless the parties agree otherwise in writing.

Section X.06. Time Limitation. Any arbitration proceeding under this Agreement must be commenced no later than one year after the controversy or claim arises. Failure timely to commence an arbitration proceeding constitutes both an absolute bar to the commencement of an arbitration proceeding with respect to the controversy or claim, and a waiver of the controversy or claim.

Section X.07. Venue. The arbitration is to be conducted in

_____.

Section X.08. Submission to Jurisdiction. Each party shall submit to any court or competent jurisdiction for purposes of enforcement of any award, order or judgment. Any award, order or judgment pursuant to arbitration is final and may be entered and enforced in any court of competent jurisdiction.

Section X.09. Costs. The costs of any arbitration, including the costs of the record or transcripts, administrative fees, attorney's fees, and any other fees shall be paid by the party determined by the arbitrators to the losing party or otherwise allocated in an equitable manner as determined by the arbitrators.

3. 其他替代性紛爭處理程序（Alternative Dispute Resolution）

　　除了訴訟與仲裁之外，還有許多替代性紛爭處理程序可以用來解決當事人間的糾紛。紛爭處理程序可以按有沒有拘束力分成許多類型，從毫無拘束力的自行協商談判、有中立第三者協助的調解、mini trial、summary jury trial，到有強制力的仲裁，以及訴訟。沒有強制力的程序，缺點是不一定能得到結果，但優點是當事人較能控制處理的結果，

而且也較能發揮智慧，達成各方都贏的局面。訴訟或仲裁大致上會有一個最終的結果，但總是要拼個輸贏，而且最終可能是各方都輸的局面。

調解就是找一個中立第三者，促進當事人各方自行談出一個適當的解決方案。Mini trial是找中立第三人當裁判，爭議各方派代表說明自己的主張。爭議各方聽過彼此的主張之後，自行談判，如果談判不成，中立第三人提出預測性的意見，說明這項糾紛如果送到法院去，結果會如何。這樣的程序可以讓爭執的各方把自己的主張試著打一遍，充分瞭解彼此的主張之後，也可以聽聽第三人的判斷意見，以評估之後還要不要訴訟，或是如何妥協。[註13]而summary jury trial則是在美國的法院中進行，由法官主持，並組成諮詢性的陪審團，陪審團員進場時會被朦在鼓裡，以為自己真的是要做判決，等到判決作出之後，才告知是參考性的判決。因此，這樣的程序更有臨場感，爭議各方也更能模擬出審判的可能結果。[註14]

締約當事人可以考量個別交易的需求，設計混搭適合的紛爭處理程序。例如用mini trial搭配仲裁應該是一個相當適合的方式，當事人找中立的第三人當裁判，一邊談判，一邊論辯，再依據第三人參考性的判斷意見決定是否修正妥協方案，如果還是無法達成妥協方案，就送仲裁。

在設計此種替代性紛爭處理程序的條款時，可以參考相關機構，例如International Institute for Conflict Prevention & Resolution[註15]的網站。

13　Stephen B. Goldberg, Frank E.A. Sander, Nancy H. Rogers & Sarah Rudolph Cole, *Dispute Resolution* 313. (5th ed. Wolters Kluwer 2007)

14　Goldberg 等前揭合著第317頁。

15　https://www.cpradr.org/, latest accessed on September 1, 2023.

債權轉讓與債務承擔條款

　　契約起承轉合，轉折部分最後一個條款就是債權轉讓（Assign-ment）與債務承擔（Delegation）。

第一節　債權轉讓與債務承擔的定義

1. 債權轉讓（Assignment）的定義

　　在美國契約法之下，債權轉讓就是轉讓人透過意思表示，將債權轉讓給他人，從而轉讓人請求履行的權利一部分或全部消滅，而受讓人則取得請求履行的權利[註1]。例如B欠A美金一百元，A將這項權利讓給C。A就不能再向B請求支付美金一百元，但C可以向B請求支付美金一百元[註2]。

　　但是，轉讓債權的同時，除非明白表示只有轉讓債權，而不移轉債務，否則債務也會被同時移轉。[註3]

2. 債務承擔（Delegation）的定義

　　債務承擔指的則是債務人將其應履行的義務委託他人履行[註4]，或者將某項給付義務的條件委託他人加以實現[註5]。所謂「將某項給

1　*Restatement (Second) of Contracts* § 317(1) (1981).

2　*Restatement (Second) of Contracts* § 317 cmt. a (1981).

3　*Restatement (Second) of Contracts* § 328 (1981). U.C.C. § 2-210(4)也有類似規定。美國統一商法典 Article 2 是專門適用於產品(goods)交易的規定，所以不適用於不動產交易、服務交易、或其他交易事件。

4　*Restatement (Second) of Contracts* § 318 (1981).

5　*Restatement (Second) of Contracts* § 319 (1981).

付義務的條件委託他人加以實現」，從契約法第二次整編（Restatement Second of Contracts）提供的例示觀察，係類似於踐成契約的狀況，亦即當事人一方完成一定條件即可請求報酬，但是並不承擔完成該條件的義務。

締約者對於債權轉讓、債務承擔是要允許、禁止，或者選擇在絕對禁止與完全不限制之間的各種不同態樣呢？在做決定之前必須先想想各種可能的需求。

第二節　當事人可能的需求

如果不能瞭解當事人的需求，任何的討論與設計都是閉門造車。所以，我們必須先想想，到底締約當事人對於自己、對於他方，在債權轉讓、債務承擔上可能有什麼樣的需求。

1. 保留己方日後活動的空間

締約時，當事人通常希望保留自己日後移轉債權或債務的可能性。例如賣方取得的應收帳款，如果能夠加以轉讓，就能減少交易成本、改善現金流、降低被倒帳的風險。至於債務移轉也是如此，有時候發現某項業務與己方的核心競爭力不符，處分轉讓該項業務或許能提升獲益，此時債權跟債務往往是隨同業務一併移轉的。

2. 債權債務專屬性的考量

相對的，當事人可能又希望限制交易對手轉讓債權的可能性。之所

以希望能限制，當然不是為限制而限制。而可能是原交易關係具有較強的專屬性，也就是看重交易對象本身的某些特質。比方說可能擔心對方將債權轉讓給像是討債公司那樣的第三人，用不當的方式來催款。或是在技術合作的交易中，合作對象把共同生產、研發的權利轉讓給我方的競爭對手，那豈不是借寇兵、齎盜糧。或是在租賃契約中，房客把房子轉租或分租給不知何方神聖，讓房東每天心驚肉跳。而在債務承擔上，新的債務人到底有沒有能力、有沒有誠信履約，自然是原權利人所關切的。

3. 擔心交易對象債權債務分離，減弱了履約的品質

當事人一方也可能擔心交易對象將債權、債務分離，造成履約品質降低。舉個例子，站在房客的立場來想，房客會擔心房東把租金債權轉讓給第三人，以後房子哪裡有什麼問題需要維修，房東可能不情不願、三催四請還不來處理。

瞭解締約者各種合理可瞭解的需求之後，我們也必須知道法律上有什麼樣的強制規定，避免擬訂的條款在訴訟後無法執行。

第三節　債權轉讓與債務承擔的法律規範

各個法域可能有保護債權轉讓的強行規定，需要特別留意。在跨國交易中如何因應此種規範差異，做最適切的條款設計是非常重要的。債權轉讓與債務承擔條款設計得不周全，往往會嚴重影響當事人權益。以下由一件發生在美國的案例談起。

案例一

訴外人T. D. J. Builders將房屋出售給訴外人Damons夫婦，並約定分期付款。^(註6)訴外人Damons夫婦後來繳不出分期付款，又積欠本案原告Francis R. Belege一筆款項，於是Damons夫婦把他們跟T. D. J. Builders簽訂的房屋買賣契約轉讓給本案原告Francis R. Belege。

但是，T. D. J. Builders與Damons夫婦簽訂的契約中有一條約定「買受人未經賣方書面同意不得轉讓本契約。」("This agreement…shall not be assigned by Purchasers without the written consent of the Seller thereto.")，於是Damons夫婦與受讓人Francis R. Belege即聯繫賣方T. D. J. Builders，徵詢其同意。T. D. J. Builders拒絕同意。

受讓人Francis R. Belege仍同意接受房屋所有權的轉讓，並向本案被告Aetna Casualty and Surety Company投保火險，也繳了保險費。之後，房子失火，原告就向被告請求保險金。被告就用原告受讓的房屋買賣契約訂有限制轉讓條款，因此原告對保險標的無保險利益拒絕理賠。

承審的紐約州法院認定解釋契約中限制轉讓條款效力的基本原則之一在於，除非契約用語明確表示任何轉讓均屬無效，否則只會解釋為締約當事人負有不轉讓的不作為義務。因此在本案中，訴外人Damons夫婦違反此項不作為義務，所以訴外人T. D. J. Builders對Damons夫婦得請求違約賠償，但不影響原告Francis R. Belege受讓的權利。

6　*Belge v. Aetna Cas. & Sur. Co.*, 39 A.D.2d 295, 334 N.Y.S.2d 185 (1972)

在美國，法院對於限制債權轉讓的條款會盡可能採取限縮解釋，因為這樣的限制妨礙了商業活動。如果契約只寫禁止或限制轉讓契約，那法院可能解釋成是限制債務承擔，但不限制債權轉讓。即使明寫禁止或限制轉讓契約下的各項權利，法院也會解釋成這只是課予當事人不轉讓債權的義務，如有違約，就是損害賠償，但債權轉讓仍屬有效。所以在跨國交易中，我們必須分析相關法域的法律規範，瞭解當事人真正的需求，並有效地落實在契約條款的設計中。

1. 債權轉讓的相關法律規則

在臺灣，債權原則上可以移轉，但依照(1)債權之性質、或(2)當事人的特別約定，或因為(3)債權禁止扣押，而不得讓與者，不在此限（註7）。所以若無特約，債權人將債權轉讓第三人，原則上根本毋庸債務人同意。只是非經原債權人或受讓人通知債務人，對債務人不生效力（註8）。

在美國，契約法第二次整編建議的規範是契約權利原則上可以移轉，但如果(1)債權移轉會顯著改變債務人的義務、或顯著增加債務人在契約上的負擔或風險、或顯著侵害債務人獲得對待給付的機會、或顯著減少對待給付對債務人的價值、或(2)法律或公共政策禁止、或(3)契約合法地禁止轉讓者(assignment is validly precluded by contract)，則除外（註9）。從「assignment is validly precluded by contract」這樣的詞句，我們可以察覺到，契約要限制轉讓，還不見得有效呢？

7　臺灣民法第294條第一項。
8　臺灣民法第297條第一項前段。
9　*Restatement (Second) of Contracts* § 317(2). U.C.C. § 2-210(2) 也有類似規定。

　　而在Common law的判例規則下，還有其他例外不能轉讓債權的情況，包括專屬性較強的服務契約（例如律師、醫師的服務）、需求承包契約（requirement contracts）、產能購買契約（output contracts）以及未來才發生的權利。

　　另外，在紐約州，除非契約明訂准許轉租，否則住宅用房屋的房客未經房東許可，不能轉租，房東可以因為任何理由拒絕同意。但相對的如果房東拒絕同意是不合理的，房客可以終止租約〔註10〕。

　　在中國，依據民法典規定〔註11〕，債權人可以將債權全部或一部分轉讓給第三人，但根據(1)債權的性質、或(2)當事人約定，或(3)法律規定，而不得轉讓者，除外。關於通知債務人的部分，亦是規定未經通知債務人，對債務人不生效力，所以是只要通知債務人，並不需要債務人同意。而且通知轉讓後即不能撤銷，除非受讓人同意撤銷。但從法條文義上看來，只能由債權人通知債務人，而不能由受讓人通知債務人〔註12〕。而兩岸同樣也有債權移轉之後，債務人可否對受讓人主張抵銷或其他抗辯的規範。

　　涉及跨國交易時，法律或公共政策如強行禁止不得轉讓者，當然就不能讓與。但契約性質不能讓與的，契約雙方能否特別約定同意讓與呢？另外，當事人特別約定不得讓與，法院會不會認為限制讓與的約定不符合公平正義，而不承認其拘束力，或者至少挑剔其用字遣詞或其他內容或格式的問題，回應訴訟當事人一方的訴求，想辦法讓這樣的約定不發生拘束效果呢？

　　有些債權，是法律規定一定可以轉讓，這類債權限制轉讓也沒用。

10　New York Real Property Law §226-b.
11　中華人民共和國民法典第545條第1項。
12　中華人民共和國合同法第79條。

例如美國統一商法典規定票據、健康保險的保險金不能限制轉讓^{（註13）}。

　　除了法律明文的限制之外，對於可以轉讓的債權，如果當事人約定限制轉讓，美國法院確實還可能對限制轉讓條款做較為限縮的解釋。法院有兩種限縮解釋的可能方法。^{（註14）}

　　第一種限縮解釋方法是盡量將限制債權轉讓條款解讀為只是「禁止移轉債務」。而這也是美國統一商法典（Uniform Commercial Code）跟契約法第二次整編所採取的立場。契約法第二次整編表示，如果當事人約定限制「轉讓契約」（assignments of the contracts），若無其他明示的意思，這只限制債務人將本來應該履行的債務、或應實現的條件，移轉由第三人承擔。而如果當事人約定「禁止轉讓契約之下的權利」（assignment of rights under the contract），則除非有相反的意思表示，否則仍不禁止當事人移轉因為違約所生的損害賠償請求權，也不禁止當事人將完全履行義務後得以享受的權利移轉第三人。而且即使債權人將債權移轉第三人，債務人也只能向債權人請求違約的損害賠償，債權移轉本身仍屬有效。^{（註15）}美國統一商法典也有類似規定^{（註16）}，不過美國統一商法典§2-210(2)對於違約所生的損害賠償請求權，以及完全履約後得以享受的權利，是完全開放轉讓，即使當事人契約明文限制也還是可以轉讓。

　　第二種限縮解釋的方法，則是解讀成當事人仍然可以轉讓契約下的權利，最多就是造成損害時，負擔賠償責任。這與臺灣民法第294條第2項規定「不得讓與之特約，不得以之對抗善意第三人」有些類似。但

13　U.C.C. §9-408. Tina L. Stark, *Assignment and Delegation*, in *Negotiating And Drafting Contract Boilerplate* 51-52 (Tina L. Stark et al. ed., ALM Publishing 2003).

14　參閱前引 Stark 著作第36-38頁。

15　*Restatement (Second) of Contracts* §322 (1981)

16　U.C.C. §2-210(2)&(3).

美國的法院甚至不談第三人是善意或惡意。

　　如果當事人想要讓與債權，只要不涉及公共政策問題，許多國家是持較開放的態度，但是基於兩點考量，建議在契約中仍然明確寫出當事人得轉讓債權。第一，如果在契約中載明得轉讓，訴訟時，另一方較難辯稱契約性質不允許轉讓（註17）。第二，中國民法通則第91條仍然規定，讓與合同的權利需要取得合同另一方的同意。雖然上述規定可能在個案中被合同法的規定取代，但在跨國交易中，我們難以透澈地瞭解各個國家的實體法律在個案訴訟中會如何被操作，所以如果想要讓與債權，寫清楚可以轉讓債權還是最安全的。

2. 債務承擔的相關法律規則

　　在中國，必須經過債權人同意，債務人才能將債務移轉由第三人承擔（註18）。

　　在臺灣，債務承擔必須經過債權人承認，才能對債權人發生效力（註19）。債權人如果拒絕承認，原債務人或承擔人得撤銷承擔的契約（註20）。

　　在美國common law的法則下，除了四類債務不能任意由債務人移轉以外，原則上債務人可以將債務移轉由第三人承擔。第一個例外是履行債務涉及債務人個人的判斷或技能的，例如表演契約，就算承擔債務的人比原債務人更會演戲，原債務人也不能自行決定將債務移轉給第三

17　*Restatement (Second) of Contracts* § 323(1)(1981)

18　中華人民共和國民法典第551條。

19　臺灣民法第301條。

20　臺灣民法第302條第2項。

人。第二種例外情況是債權人與債務人之間有特殊的信賴關係，像是當事人與律師之間。第三種例外則是移轉債務會顯著變動債權人原先對履約的期待，像是需求承包契約、或產能購買契約。如果債務人A廠本來每個月生產一百件指定產品，全部銷給債權人B，債務人將這項完全供應契約的供應義務轉由第三人C廠履行，C廠每月生產一千件指定商品，當然會嚴重變動債權人B的預期。最後一種例外，則是契約限定不能移轉債務的情況。限制債務承擔條款通常獲得法院較大尊重。

在美國，雖然債務人可以任意將債務轉由第三人承擔，但是，原債務人移轉債務之後，仍對債權人負擔責任[註21]。如果債權人明白同意將債務轉由第三人承擔時，則原債務人可以免責，這種情況被視為是契約更新（novation）。講白話，就是重新締約，讓原債務人退出契約，讓第三人加入契約。但債權人不同意，並不影響債務承擔契約的效力，除非債務承擔契約將債權人的拒絕同意列為解除事由，或者將債權人的同意列為停止條件、或不同意列為解除條件。

如果未約定限制債務承擔，僅僅依賴債務的屬性不見得能論證得出限制債務承擔的效果。契約法第二次整編第三百一十八條第三項表示，除非當事人另行約定，否則就算指定特定人履約，必須特定人的履約或控管整個活動對債權人有重大的利益，才不能再轉包出去[註22]。

從前述說明，我們可以得知，即使是債務承擔，若要限制還是在契約明文約定為宜。相對的，如果要保留將債務移轉第三人承擔的機會，也是以明定為宜，避免對方日後爭執相關債務的屬性不適宜轉移[註23]。

---

21 *Restatement (Second) of Contracts* § 318(3) (1981). U.C.C. § 2-210(1)

22 U.C.C. § 2-210(1)也有類似規定。

23 *Restatement (Second) of Contracts* § 323(1) (1981).

第四節　條款設計方法

1. 考量交易專屬性及其他因素而限制移轉

　　如前所述，美國法院對於限制債權轉讓的條款可能做嚴格解釋，或者限縮字面的意義，變成只限制移轉債務，而不限制轉讓債權，或者讓債權轉讓變成違約但仍然有效。因此，不能只是課予締約當事人未經同意不得移轉債權、債務的義務，還要直接寫明移轉無效。為了避免法院認定這樣的約定不能對抗善意第三人、甚至也不能對抗惡意的第三人，最好再加上終止或解除契約條款（註24）。不過，在整個契約編排上，終止事由也可以集中到契約終止條款，而不列在權義轉讓條款中。當事人一方違約移轉債權債務時，另一方就可以終止或解除契約。此外，預定違約移轉債權債務的損害賠償數額（註25），並要求被限制的一方確認存在假處分的條件，也都至少有利於造成道德上的壓力。

　　美國統一商法典也提供了一項對應債務移轉的安全機制，也就是要求提供履約擔保（註26），因此，我們不要等到事情發生以後再來要擔保，可以在契約中事先寫明，把提供特定金額的人保、物保或銀行保，列為同意權人考量是否同意移轉的因素之一。

2. 為避免債權債務分離，造成履約品質低落而限制移轉

　　若是擔心債權債務分離會造成履約品質低落，可以限制債權債務的

24　在英文中其實 "terminate" 可能指終止，也可能指解除。並未如同臺灣那樣有明確的區別。
25　在美國，違約原則上不能課予懲罰性損害賠償。
26　U.C.C. § 2-210(5)

移轉必須合併進行。不過放任對方自行判斷移轉的條件，仍有風險。而且債權、債務的移轉可能因為特殊情況，嗣後發現其中一項出現問題，以致於仍然曝露在風險中。而且債權債務仍可能被移轉到履約條件不佳的第三人。因此，仍宜直接採取未經同意、不得移轉的限制模式，只是寫明在是否同意許可移轉債權債務的過程中，合併移轉是考量事項之一。

3. 同意債權轉讓、債務承擔時的設計方法

締約當事人可以直接了當的禁止轉讓債權或債務承擔，但禁止債權轉讓條款的合理性可能被質疑。如果是約定必須經過他方同意才能轉讓債權，則這種條款的合理性較容易被肯定，但個案拒絕同意的理由仍需斟酌，法院可能會介入審查，就如案例一的情況，或者在設計契約時就預先想清楚可能因為什麼理由必須拒絕，而需要為這些理由預先思考正當性，並在契約中適當地列出。

此外也必須留意，在美國，如果是不動產租賃契約，出租人同意轉租一次，之後原則上，新承租人就是諸法皆空、自由自在，可以任意轉租了，因為，出租人被推定已經豁免（waive）新的承租人日後再次轉讓時需要取得出租人同意的義務，除非出租人在第一次同意轉租時明白表示以後再轉租還是必須事先取得同意。因此，為避免當事人一方的某一次同意被推定為豁免日後取得同意後再轉讓債權的義務，最好在限制條款中寫明一次同意不代表豁免。同時，在同意移轉的書面中也要載明此項同意是一次性的，日後再移轉必須再次事先取得同意。

如果要求必須我方同意，對方才能移轉債權、債務，對方很可能在法庭訴訟上抗辯我方的拒絕同意是不合理的。如果我方有較多的談判籌

碼，可以在契約條款中寫明是否同意，完全由同意權人裁量判斷。即使如此寫，仍不見得完全能獨斷獨行，在大陸法系，要求誠實信用的法律不勝枚舉（註27），在英美法系，法院可能還是會在個案判斷中要求誠實信用（註28），法院即可能由誠信原則推導出不能不合理地拒絕同意。不過，美國法院也有認定當事人可以合意排除商業合理性原則（commercial-reasonableness standard）的案件（註29）。

　　基於前述考量，如果希望限制對方移轉債權債務，在契約中可以寫明，「由於本契約重視交易對象的財務狀況、履約能力、及交易當事人之間彼此適應的能力、債權債務移轉可能造成有同意權一方的重大損害，因此是否同意移轉債權、債務完全由有同意權的一方裁量判斷。有同意權的一方在考量是否同意移轉時可能考量的因素包括移轉對象的能力、營業性質、債權債務是否合併移轉、市場情勢、同意權一方的營業規劃、移轉人或受讓人是否提供適切擔保及有同意權一方認為相關的其他事項。」此處的重點在於最好適當交代給予有同意權一方絕對的裁量權限的理由，以便日後能通過個案爭訟程序中法官或仲裁人的審查，當然理由要列得詳細、明確，還是延用相關法條中較為模糊的詞語，可以個案斟酌。而列出同意與否可能考量事項的用意在於列出各種可能的理由，即使這些理由與轉讓人自身的行為沒有太大關聯（例如市場情勢），如果對方簽字認帳，事後對方要再爭執則難度較高。

27　參閱臺灣民法第一百四十八條、中華人民共和國民法通則第四條、合同法第六十條，

28　參閱前引 Stark 著作第 67-69 頁。U.C.C. § 1-203.

29　*Shoney's LLC v. MAC East, LLC*, No. 1071465 (Ala. Jul. 31, 2009)("[w]here the parties to a contract use language that is inconsistent with a commercial-reasonableness standard, the terms of such contract will not be altered by an implied covenant of good faith. Therefore, an unqualified express standard such as 'sole discretion' is also to be construed as written.").

4. 意料之外的債權轉讓或債務承擔

設定質權也可能產生債權轉讓的效果，所以在契約定義上宜將質權設定也列入。

另外，在契約關係下，除了締約當事人一方跟第三人合意將特定債權直接轉讓給第三人、或合意設定權利質權、或將債務移轉由第三人承擔之外，還有其他可能的事件可能造成債權轉讓或債務承擔的效果。(註30)也就是法定的債權移轉或債務承擔，例如信託、破產管理程序、繼承，以及企業合併或出售股份，或者甚至締約一方當事人死亡。對於這些事件的效果，契約當事人能否預先禁止發生權義移轉效果？

由美國法院的判決進行分析，締約當事人如果未明文約定限制這些法定債權轉讓的效力，法院通常會認定當事人可以透過合併或法定債權移轉的方式轉讓其債權。但在Nicholas M. Salgo Associate v. Continental Illinois Properties(註31)此一案件中，法院則認定即使當事人契約中的文字不是那麼明確，但其限制債權轉讓條款的範圍也包括透過企業合併所為的債權轉讓。因此，如果當事人在契約中約定得相當明確，美國法院確實有可能肯認此種限制轉讓條款的效力。但這畢竟受限於法院個案的解釋

所以如果已經打定主意，無論如何不能轉讓任何債權或移轉任何債務，或許可以設計成契約當事人一方死亡、解散或喪失行為能力，他方可以終止契約。

但如果保留可以移轉債權債務的空間，或者第三人透過其他方式，例如繼承、公司併購或其他方式承受契約權利義務的可能性，就需要約

---

30 參閱前註2引Stark著作第42-50頁。
31 532 F. Supp. 279 (D.D.C. 1981)轉引自前註2引Stark著作第48頁。

定契約對承受者效力的條款（successors and assigns）。此條款可以單獨作為一條，但也可以加入債權轉讓與債務承擔條款中附帶說明。

撰擬契約對承受者效力條款時，需注意在「successors and assigns」之前加上「permited」，以避免當事人一方未經許可就將契約移轉給第三人，然後回頭主張因為寫了契約對承受者效力條款，所以默示允許轉讓。契約對承受者效力條款究竟有無默示許可轉讓的意思，美國各法院的見解相當紛歧。因此，若絕對不允許轉讓，就不要寫契約對承受者效力條款以避免不必要的困擾，另外加上

「permitted」表示並非可以隨便移轉的[註32]。

5. 範例條款

（一）各方均受限制

> **例句一：**
>
> --
>
> **Article　　Assignments.**
>
> **(a) No Assignment or Delegation.** No Party may assign any right or delegate any duty under this Agreement without prior written consent of the other Party.
>
> **(b) Consent.** The Party with consent right shall not withhold its consent unreasonably.
>
> **(c) Non-Waiver.** Any Party's consent for specific assignment or delegation is not a waiver of consent for any future

---

32　參閱前註3引Stark著作第81-93頁。

assignment or delegation.[註33].

(d) Void Assignment or Delegation. Any purported assignment or delegation violating subsection (a) is void.

(e) Definition of Assignment. Assignment means any transfer, pledge or any other form of alienation or encumbrance of rights or interest under this Agreement made by any Party for any Nonparty, with or without consideration, voluntarily or involuntarily, whether by merger, consolidation, dissolution, operation of law, or any other manner.

(f) Specific Assignments Prohibited. Assignment prohibited under subsection (a) includes, without limitation, assignment

(i) of any claim for damages arising out of the other Party's breach of the

whole contract;

(ii) of any rights arising out of the assignor's full performance of this Agreement;

(iii) by change of control; and

(iv) by merger no matter that Party is surviving or disappearing corporation.

(g) Liquidated Damages. If any Party assigns any rights or delegates any duties violating subsection (a), that violating Party shall pay the other Party [specific amount or other calculation]

---

33 *See* Milton R. Firedman, *Friedman on Leases*, § 7:3.3 (5th ed. PLI, 2004), Tina L. Stark, *Assignment and Delegation*, in *Negotiating And Drafting Contract Boilerplate* 55,58, 63,78 (Tina L. Stark et al. ed., ALM Publishing 2003), 72 *Del. Laws,* c. 303, § 1(r), 75 *Del. Laws,* c. 148, § 1(r), Evelyn C. Arkebauer, *Cumulative Remedies And Election of Remedies*, in *Negotiating And Drafting Contract Boilerplate* 229 (Tina L. Stark et al. ed., ALM Publishing 2003), *Restatement (Second) of Contracts* § § 317(2)(a), 318(2) (1981), and Marcel Fontaine & Filip De Ly, *Drafting International Contracts: An Analysis of Contract Clauses* 552 (Transnational Publishers, Inc. 2006)

as liquidated damages, which is not a penalty.

(h) **Injunctive Relief.** If any Party assigns any right or delegates any duties under this Agreement without the other Party's prior written consent, it will cause the other Party irreparable damage for which monetary remedies is inadequate. The other Party may obtain injunctive relief to restrain the assignment without requirement of a bond or proof of monetary damage or an inadequate remedy at law, in addition to all other remedies available at law or equity.

(i) **Termination for Unconsented Assignment.** If any Party assigns any right or delegates any duty violating subsection (a), the other Party may terminate this Agreement, and such termination is effective as of the prohibited assignment's or delegation's occurrence. Any such termination does not affect the other Party's claim for damages.

(j) **Successors and Assigns.** This Agreement binds and benefits the Parties and their respective permitted successors and assigns.^{（註34）}

（二）一方受限、他方不限制

例句二：

Article　Assignments.

(a) **No Assignment by [One Party's Name].** Without prior

34　參閱前引 Stark 著作第 87 頁。

written consent of [the Other Party's Name], [One Party's Name] shall not assign following rights under this Agreement to any Nonparty other than Affiliates of [One Party's Name] described in Exhibit A:

(i) Any claim for damages arising out of the other [Name of the other

Party]'s breach of the whole contract.

(ii) Any rights arising out of the assignor's full performance of this Agreement.

(b) No Delegation by [One Party's Name] [One Party's Name] shall not delegate any duty under this Agreement without prior written consent of [the Other Party's Name].

(c) Impact of Assignment. Assignment of rights under this Agreement by [One Party's Name] will materially

(i) change the duty of [the Other Party's Name];

(ii) increase the burden and risk imposed on [the Other Party's Name];

(iii) impair [the Other Party's Name]'s chance of obtaining return performance; and

(iv) reduce value of return performance to [the Other Party's Name].

(d) Impact of Delegation. [The Other Party's Name] has substantial interesting in having [One Party's Name] perform and control the acts promised.

(e) Consent. Due to the facts stated in subsection (b) and (c), [the Other Party's Name] may consider following factors and decide whether to consent at its sole discretion:

(i) Capability of assignee or delegate.

(ii) Business nature of assignee or delegate.

(iii) Market situation.

(iv) Business plan of [the Other Party's Name].

(v) Concurrent Assumption of duties by assignee.

(vi) Adequateness of surety or assurance provided by assignor, assignee, delegator or delegate..

(f) Non-Waiver. [The other Party's Name]'s consent for specific assignment is not a waiver of consent for any future assignment or delegation.

(g) Void Assignments. Any purported assignment or delegation violating subsection (a) or (b) is void.

(h) Definition of Assignment. Assignment means any transfer, pledge or any other form of alienation or encumbrance of rights or interest under this Agreement made by any Party for any Nonparty, with or without consideration, wholly or partially, voluntarily or involuntarily, whether by merger, consolidation, dissolution, operation of law, or any other manner..

(i) Specific Assignments Prohibited. Assignment prohibited under subsection (a) includes, without limitation, assignment

(i) by change of control; and

(ii) by merger no matter [One Party's Name] is surviving or disappearing corporation.

(j) Liquidated Damages. If [One Party's Name] assigns any right or delegates any duty violating subsection (a) or (b), [One Party's Name] shall pay [The other Party's Name]　　[specific amount or other calculation] as liquidated damages, which is not a penalty.

(k) Injunctive Relief. If [One Party's Name] assigns any right or delegates any duty violating subsection (a) or (b), it will cause [Name of the Other Party] irreparable damage for which monetary remedies is inadequate. [Name of the Other Party] may obtain injunctive relief to restrain the assignment without requirement of a bond or proof of monetary damage or an inadequate remedy at law, in addition to all other remedies available at law or equity.

(l) Termination for Unconsented Assignment. If [One Party's Name] assigns any right or delegates any duty violating subsection (a) or (b), [the Other Party's Name] may terminate this Agreement, and such termination is effective as of the prohibited assignment's or delegation's occurrence. Any such termination does not affect the other [Name of the Other Party]'s claim for damages.

(m) Assignment or Delegation by [the other Party's Name]. [The other Party's Name] may assign any right or delegate any duty under this Agreement without prior written consent of [One Party's Name].

(n) Successors and Assigns. This Agreement binds and benefits the Parties and their respective permitted successors and assigns.

（三）雙方均不限制

例句三：

Article 　Assignments.

(a) Assignment or Delegation. Any Party may assign any right or delegate any duty under this Agreement without prior written consent of the other Party.

(b) Successors and Assigns. This Agreement binds and benefits the Parties and their respective permitted successors and assigns.

第五節　談判過程的折衝與妥協

締約過程往往不是一方說要怎樣就怎樣，而是不斷的折衝與妥協。在這過程中，必須審慎思考我方與對方真正的利益何在？真正擔憂的事情為何？設定談判的策略，有些點可以退讓、有些點必須堅持。必須運用創意來達成雙贏的局面。

舉例來說，契約關係之下有一籃子的權利義務，是否全部都限制移轉？還是有些可以開放、有些必須限制。可以做更細膩的處理。

此外，有些第三人可以列入承受債權債務的白名單，例如締約當事人的關係企業。不過，最好是列明清楚哪些關係企業（註35），而且如果開放移轉給關係企業，關係企業的股權變動也必須納入管控，也就是再

---

35 參閱前引 Stark 著作第 70-71 頁。

303

納入移轉條款的射程範圍。

第六節　重點回顧

綜合前述各種考量，包括債權轉讓可能性的預設規則，觸發債權轉讓的事件、可運用的管控手段、以及談判妥協的需求，我們可以整理出幾點注意事項，在撰寫契約中債權轉讓條款時，必須加以留意。

1. 如要保留我方轉讓債權的可能性，最好寫明。

2. 如要限制對方轉讓債權，必須寫明是限制轉讓契約之下的權利（assignment of rights under contract），而非只寫限制契約轉讓（assignment of contract）。因為只寫限制轉讓契約，可能被解讀為僅限制債務承擔，而不限制債權轉讓。

3. 觸發債權轉讓的事件不只有意定的債權轉讓（意定轉讓不只有直接移轉，還有質權設定的情況），還包括法定的債權轉讓，以及企業併購的狀況，必須納入移轉的定義。

4. 將法律預定但不強制開放移轉的項目也特別寫入禁止移轉的清單。

5. 如要保留我方轉移債務的可能性，最好寫明。如要限制對方移轉債務，也要寫明。

6. 限制移轉債權、債務的方法建議用明文禁止，例外經事先書面同意可移轉的方式處理。

7. 用五道大鎖防堵移轉，包括(1)設定不作為義務、(2)違約移轉的賠償、(3)約定未經同意的移轉無效、(4)違約移轉時，他方得終止解除

契約以及 (5) 違約移轉時，他方得提起假處分，違約方不抗辯假處分之適切性。

8. 書面同意要加上保留條款，不能疏忽而變成豁免徵得同意的義務（waiver）。

9. 同意與否完全由同意權人裁量判斷，但最好交代給予同意權人此種裁量空間的理由。

10. 列出同意權人決定是否同意時可能考量的各種因素，並可要求第三人提供擔保作為同意的前提條件。

11. 可區別開放移轉的標的與限制移轉的標的，以利達成折衷方案。

12. 可列明開放移轉的對象的白名單、或限制移轉的對象黑名單，以利達成折衷方案。

各種結尾條款與簽名部分

　　談完了契約的起、承與轉之後，我們要談契約最後的合部，也就是結尾部份，也有組合稱為雜項條款、其他條款或一般條款。契約的結尾部分通常包括通知條款、效力可分、完整合意、多語條款、正本份數以及簽名頁。

　　本章將先討論通知條款，通知條款看似平淡無奇，但筆者也遇過交易對象營運不善，終止通知被拒收退件，究竟是否有效送達的疑難情況，因此通知條款也有它的重要性。

第一節　通知條款（Notice）

　　通知條款要規範的事項包括通知的方法、收受通知的人員、聯繫資料與通知生效的時點。許多人可能忽略了通知條款的重要性，通知條款有兩項重要的影響。首先是通知生效時點的重要性。在英美法系，要約意思表示生效時點與大陸法系一致，採到達主義，要送到收信人所在地才生效，而承諾的意思表示生效時點通常是採用發信主義，也就是說當事人把通知裝到信封中，交郵局寄送時，就生效了，英美法的這項規則也稱為郵箱規則（Mail Box Rule）。而在大陸法系，承諾的意思表示仍採用到達主義，所以訊息要送到收信人那裡，在收信人控制之下，可以拆信加以閱讀，才算生效。第二個重要性在於，美國的訴訟文書是要當事人自己送達到對造手上的，約定通知方法可以避免送錯地方，送達不到。

　　通知條款同樣地也可以分成通用的（generic）通知條款，與特別情況適用的（specific）通知條款兩個類型。(註1)通用的通知條款指的是適

1　Steven R. Berger, *Notices* in *Negotiating And Drafting Contract Boilerplate* 467-468 (Tina L. Stark et al. ed., ALM Publishing 2003).

用於各種需要通知的情況，不管是下單、接受訂單、付款請求、違約通知，除契約另有明文排除之外，均一體適用。而特別情況適用的通知條款，就是在契約中提到特定通知事項時，特別說明通知應如何發送，向誰發送、何時生效等問題。例如在借款契約中提到銀行宣告違約時，應該如何通知借款人、通知應寄到什麼地點，通知何時生效。以下分別說明。

1. 通用的通知條款

通用的通知條款有幾個構成的部分，包括(1)通知的定義、(2)通知是給付義務或是主張權利的條件？(3)通知的方式、是否需要簽章、收信人與地址、電話、電子信箱、(4)通知生效時點。

（一）通知的定義

在通用的通知條款中，難以對通知做出很確切的定義，一般只能用「notice」、「demand」、「request」、「communication」這類近似字詞大致描述。舉例來說：

> **例句一：**（註2）
>
> Any party giving or making any notice, request, demand or other communication (each a "Notice") pursuant to this Agreement …….

2　Steven R. Berger 前揭著第 468 頁。

（二）以通知為給付義務，或權利之前提條件

在撰寫通用的通知條款時，我們沒辦法預知日後真的發通知是契約規定必須發通知（作為一種義務，不履行有賠償責任），還是作為主張權利的前提條件（例如要開請款單，才能請款，但不開請款單最多就是要不到錢，不用再賠錢）。因此，我們必須用彈性的方式改寫例句一，說明通知可能是作為義務，也可能作為主張權利的條件。

例句二：

When any party is obliged to or for asserting rights must give or make any notice, request, demand or other communication (each a "Notice") pursuant to this Agreement ......

（三）通知的方式

通知的方式可能是口頭或書面，口頭必須有聯絡電話及聯繫人，書面也必須有地址、電郵信箱、傳真方式的指定以及收信人。透過郵局或快遞寄送也可以再指定必須是掛號或可用平信。此外是否加密或用電子簽章，也可以特別說明。特別提醒，不宜過度僵化地限定只能用掛號附回執的方式寄送，避免送不到造成困擾，因為並非全球各地都有通用的掛號附回執服務，例如臺灣的郵局寄送地也是有限的，在臺灣要用國外的郵政服務也是很困擾的，所以跨國通訊最好是指定用可靠有聲譽的快遞即可。

例句三：^{（註3）}

Methods of Notification. When any party is obliged to or for asserting rights must give or make any notice, request, demand or other communication (each a "Notice") pursuant to this Agreement, the Notice is to be given (a)in writing signed by the sender and (b) by using one of the following methods:

(i) personal delivery,

(ii) Registered or Certified Mail, in each case, return receipt requested and postage

prepaid, or nationally/internationally recognized overnight courier, with all fees

prepaid,

(iii) facsimile or telex, or

(iv) email.

Addresses and Addressees.

Company Name:

Mailing Address

Courier Address (if different)

Attention

Facsimile Number

Telephone Number

Email

3　參考 Steven R. Berger 前揭著第 471-473，486 頁並加以改編。

（四）通知生效時點

通知的生效時點可以約定為發訊人送出之時，或者是收訊人收到之時。但無論是約定送出之時或收到之時，隨著傳訊媒體的差異，具體的時點仍有必要做精準的約定。例如：

例句四：（註4）

Effectiveness of a Notice. Except as provided otherwise in this Agreement, a Notice is effective only if the sender has complied with subsection _and if the receiver has received the Notice. A notice is received,

(i) if a Notice is delivered in person, or sent by Registered or Certified Mail, or nationally/ internationally recognized overnight courier, upon receipt as indicated by the date on the signed receipt.

(ii) if a Notice is sent by facsimile, upon receipt by sender of an acknowledgement or transmission report generated by the machine from which the facsimile was sent indicating that the facsimile was sent in its entirety to the receiver's facsimile number.

(iii) if a Notice is sent by telex, upon confirmed answerback.

(iv) if a Notice is sent by email, as the sender receive confirmation from the receiver.

Or

(v) if the receiver rejects or otherwise refuses to accept the

---

4　參照改編Steven R. Berger前揭著第476頁。

Notice, or if the Notice cannot be delivered because of a change in contact information for which no Notice was given, then upon rejection, refusal or inability to deliver.

2. 特別情況適用的通知條款

特別情況適用的通知條款的內容大致上與通用的通知條款相近，只是在編排上緊接著其他約定事項的條款，在通知事項的界定上更為明確，例如借款人在額度內通知增貸、或出賣人通知買受人購貨明細並請求付款，或當事人一方通知他方提前終止或解除契約。若有多數通知事項，就必須逐一訂定通知條款。

特別情況適用的通知條款必須明確說明通知是義務，還是行使權利的條件，義務就用「shall」來表示，行使權利的條件就用「must」或「may …only if」來表現。

以下提供一個例子：

例句五：（註5）

4. CREATION OF SECURITY INTEREST

4.1 Grant of Security Interest. Borrower hereby grants Bank, to secure the payment and performance in full of all of the Obligations, a continuing security interest in, and pledges to Bank, the Collateral, wherever located, whether now owned or

---

5　Loan and Security Agreement between Silicon Valley Bank and MOBITV INC., http://www.sec.gov/Archives/edgar/data/1380124/000119312511355373/d217595dex1018.htm, accessed on 2012/1/2。

> hereafter acquired or arising, and all proceeds and products thereof. Borrower represents, warrants, and covenants that the security interest granted herein shall be and shall at all times continue to be a first priority perfected security interest in the Collateral subject only to Permitted Liens. If Borrower shall at any time acquire a commercial tort claim, Borrower shall promptly notify Bank in a writing signed by Borrower of the general details thereof and grant to Bank in such writing a security interest therein and in the proceeds thereof, all upon the terms of this Agreement, with such writing to be in form and substance satisfactory to Bank.

第二節　效力可分條款（Severability）

　　臺灣民法第一百十一條規定：「法律行為之一部分無效者，全部皆為無效。但除去該部分亦可成立者，則其他部分，仍為有效。」，消費者保護法第十六條規定：「定型化契約中之定型化契約條款，全部或一部無效或不構成契約內容之一部者，除去該部分，契約亦可成立者，該契約之其他部分，仍為有效。但對當事人之一方顯失公平者，該契約全部無效。」

　　在美國，若違法或無效的條款與契約中其他條款間的關係是相互獨立，可明確區別，而且當事人間相互的給付內容可以切割，例如屋主找水管工人買材料來修水管，但依據當地法規，水管工人要有政府許可證，若水管工人未獲許可，則修水管的部分無法履行，但買材料的部分仍屬可履行，而且屋主支付的對價也可以明確區別契約。則在此種情況

下，法院會剔除契約中違法或無效的條款，調整雙方對價。

　　若契約給付內容無法切割，但違法或無效的條款在當事人合意的交易內容中並不是重要的部分，則法院會排除違法或無效的部分，但其他部分仍屬有效。然則，無論基於給付可分，或基於無效部分不具重要性而主張契約其他部分仍有效力，都必須主張者沒有嚴重的不當作為造成契約部分條款違法或無效。此外有些法院堅持只能刪減條款，不能修改條款，也就是所謂的藍筆規則。(註6)

　　以上僅是臺灣與美國的相關法律考察，若涉及其他法域，則必須重新瞭解其適用的法令。因此，個別契約如果有部分條款欠缺合法效力時，究竟會不會影響契約其他部分的效力，可能因為不同法域，法官或仲裁人的理解不同而難以預期。

　　締約者主觀上可能認為某些條款是契約的重心，若這些條款不具效力，則整個契約不能達到原先的期望。相對的，某些條款可能被認為並非至關緊要，即使無效，也不應妨礙整個契約的有效。所以，為避免交易對手或法院有不同的認定，產生意料之外的結果，有必要在契約中約定清楚。

　　該如何約定清楚呢？如前所述，我們預期某些條款是無關緊要，另些條款則是契約的核心，若核心條款不存在了，整個交易就不做了。因此，我們不能寫每一個條款都是無關緊要，也不能寫每個條款都是至關緊要，較妥當的做法就是約定原則上每一條款都是可分的，不影響其他條款的效力，但同時明訂例外必須有效，否則整個契約就失效的重要條款。例如：

6　Steven R. Berger 前揭著第 542-546 頁。

例句六： (註7)

If any provision of this Agreement is invalid, illegal or unenforceable, the remaining provisions of this Agreement remain in full force and effect, if

(i) the essential terms and conditions of this Agreement for both parties remain valid, legal and enforceable. Section _____, and _____ are essential terms and conditions of this Agreement, and

(ii) both the economic and legal substance of the transactions this Agreement contemplates are not affected in any manner materially adverse to any party.

第三節　完整合意條款（Entire Agreement）

　　完整合意條款的法理依據是口頭證據排除法則（parol evidence rule）。此處所謂的口頭證據排除，其實並不僅僅排除口頭證據，其核心的意義在於，只要在契約簽署之前或同時所合意的事項，無論是口頭合意或書面合意，而未被納入眼前簽署的這份契約，且跟眼前這份契約中意思清楚明確的條款相衝突的，則其效力會被排除。因此口頭證據排除法則要能適用有幾個前提，(1)書面契約的條款已構成最終而明確的契約，若書面契約內容曖昧不明，則當然可以主張其他合意內容以補充解釋。(2)當事人提出的相反合意跟書面契約條款在意思上相衝突而不能併存。(3)當事人提出的相反合意是簽訂書面契約前或其同時所達成

7　Steven R. Berger 前揭著第 552-553 頁。

的。

　　若書面契約簽署之前或同時所做的合意，與書面契約內容並不衝突，則是否仍被排除就取決於當事人如何約定。若當事人約定其書面契約構成完整而排他的合意，則即使不衝突的補充約定也不能主張，若當事人並未約定其書面契約構成完整而排他的契約，則當事人可主張不衝突的補充合意內容[註8]。

　　英美法系採取口頭證據排除法則，但相對的，大陸法系的契約法往往強調不拘泥於文字，而必須探求當事人真意，所以當事人可能提出相衝突的合意資料做不同主張。在跨國交易中必須確認清楚當事人究竟想要適用哪一種法則，是任何其他書面或口同的資訊交換都有效，還是只有寫在書面契約裡的條款有效，或是與書面契約條款衝突的無效，但補充書面契約條款的有效。以下舉出最強烈的完整排他合意為作為示例。若要適度允許較早或同時產生的其他合意資料以解釋、補充、甚至排除最終書面契約條款，則可自行修改例句。不過允許當事人以較早或同時產生的其他合意資料排除最終書面契約，是一種自相矛盾的作法，容易引發糾紛。

例句七：[註9]

　　This Agreement constitutes the final agreement between the parties. It is the complete and exclusive expression of the parties' agreements on the matters contained in this Agreement. All prior and contemporaneous negotiations and agreements

---

8　Ronald B. Risdon & William A. Escobar, *Merger in Negotiating And Drafting Contract Boilerplate* 562-564. (Tina L. Stark et al. ed., ALM Publishing 2003).

9　Ronald B.Risdon & William A. Escobar 前揭著第573頁。

between the parties on the matters contained in this Agreement are expressly merged into and superseded by this Agreement. The provisions of this Agreement may not be explained, supplemented, or qualified through evidence of trade usage or a prior course of dealings. In entering into this Agreement, neither party has relied upon any statement, representation, warranty or agreement of the other party except for those expressly contained in this Agreement. There are no conditions precedent to the effectiveness of this Agreement other than those expressly stated in this Agreement.

Nonetheless, the parties agree the agreements listed below are still effective

(i)

(ii)

當事人也可以明確列出其他仍屬有效的合意資料,不過仍需考量這些合意資料與最終的契約之間有無矛盾。

第四節　語言條款（Language）

契約如果同時使用多種語言作成,有時多語內容難免有不一致之處,即需約定以哪一個語言的版本為準。以下提供例句:

例句八：

This Agreement is made both in English and Mandarin.

> When English and Mandarin versions conflict, the English
> version shall prevail.

第五節 正本副本條款（Counterparts）

契約時常會有多份正副本，兩份就是duplicate，三份就是tri-plicate、四份就是quadruplicate、五份就是quintuplicate、六份就是sextuplicate、七份是septuplicate、八份是octuplicate。在古老的時候，可能是一份契約撕成兩半，雙方各留一半為憑，甚至必須以契約的交付（delivery）作為生效要件，而不僅僅是契約的簽署（execu-tion）。當然現在多半不要求以交付為生效要件，純粹只作為憑證。而契約正副本的安排還可以便利分隔數地的當事人以較省事的方式各簽一份，交給對方以表示合意，而無須大家聚集在一處簽名，或一份契約傳來傳去輪流簽名，不過在這種情況下，只簽名可能還不生效，必須將簽名的文件交給對方。

但要特別提醒，契約成立的重點在於意思表示，所以如果不是當場同時簽字，而是各在一方簽署，簽署後務必將簽署後的正本或掃描本以郵件、電郵或通訊軟體寄送給對方，這才是意思表示，如果簽完就塞到抽屜裡，對方到時候可能爭執契約並未成立。

以下例示其寫法：

例句九： (註10)

The parties may execute this Agreement in multiple counterparts, each of which constitute an original, and all of which, collectively, constitute only one agreement. The signatures of all of the parties need not appear on the same counterpart, and the delivery of an executed counterpart signature page by facsimile is as effective as executing and delivering this Agreement in the presence of the other parties to this Agreement. This Agreement is effective upon delivery of one executed counterpart to the other parties. In proving this Agreement, a party must produce or account only for the executed counterpart of the party to be charged.

若不以交付簽署後的契約為生效與權利主張的條件，則可約定：

例句十：

The parties may execute this Agreement in multiple counterparts, each of which constitute an original, and all of which, collectively, constitute only one agreement. Delivery of executed counterparts does not affect the effectiveness of this Agreement.

---

10 Frances Kulka Browne, *Counterparts* in *Negotiating And Drafting Contract Boilerplate* 585. (Tina L. Stark et al. ed., ALM Publishing 2003)

第六節　英文契約的簽名頁、附件與增補方式

1. 簽名頁

（一）結尾文字

在契約中寫完各個條款之後，當然也可以直接做簽名欄讓當事人畫押。但一般會再加上一段結尾文字，就像戲劇裡常有的「過場」，會較為平順。結尾文字的寫法很多，要點在於當事人簽名並交付契約文件。之所以要求交付文件，有部分原因是因為蓋印契約、票據、證權證券等以交付為生效要件，還有另一層考量在於確認已為意思表示，而不是簽了字直接塞到抽屜裡，所以在簽名欄讓當事人一併確認。

結尾文字可以連結到前言所載明的契約日期，當然如果前言不說明日期，而是在簽名欄註記日期，就改成「on the date opposite that party's signature」[註11]。

（二）簽名欄的設計

通常的做法是當事人稱謂（例如買方、賣方）的所有字母都用大寫粗體表示則較為醒目，當事人的名稱（某某公司）則所有字母大寫，但不粗體。實際簽名者的姓名只要第一個字母大寫，並且不需粗體。

如果當事人是法人，要先查詢相關登記網站或註冊資料確認法人註冊的名稱，尤其「Inc.」、「Corp.」、「Co. Ltd.」這些公司組織形態的簡寫不能弄錯，兩家公司可能名字一樣，組織型態不同。

11　Kenneth A. Adams, *A Manual of Style for Contract Drafting* 126 Sample 10(3rd ed., ABA, 2013).

其次，必須要求當事人提供公司章程或董事會規章以確定被授權簽名的人是誰？有幾位？並取得當事人發出的被授權人服務證明，說明被授權簽名的人確實在職，且獲授權簽名。服務證明中也需要附上被授權人的簽名，才能比對。如果當事人是美國的股份有限公司，服務證明上必須有董事會秘書（Corporate Secretary）[註12]的簽名，然後還要另外一位高階主管簽名確認董事會秘書目前確實在職。

當事人如果是合夥事業，原則上由負無限責任的合夥人（general partners）簽名。如果是美國的有限公司（limited liability companies），原則上由董事簽名。

被授權簽名者簽名處前面必須加「By:」、或「Per:」表示簽名的人是被授權簽名，而不是以自己的名義簽名。實際簽名者的職銜也要列出來。這點很重要，沒有加「By:」或「Per:」與職銜，簽名的人可能自己也要負責[註13]。範例中的「By its duly authorized signatory」只是提醒簽名的人，如果沒有授權，就不要亂簽。在契約本文中必須再加入授權條款以處理授權問題。

有時候公司被授權簽名的人是法人而不是自然人，就像法人也可以擔任董事監察人一樣。此時就必須有兩層的授權，如同範例所示[註14]。

（三）防偽機制

盡可能將契約的最後一條跟簽名欄放在同一頁。例如可以在契約的

---

12 這是高階主管，不是臺灣一般所說的秘書。
13 參閱 Stark 前揭著第 190-194 頁。"by"、"per" 是 Restatement (Second) of Agency § 156 Comment A 所建議的方法。
14 參閱 Adams 前揭著第 122 頁。

倒數第二條結束後，就另起一頁加入最後一條跟簽名欄。契約倒數第二條之後的空白處，可以插入「**INTENTIONALLY LEFT BLANK**」。如果簽名的人太多跟契約最後一條擠不進同一頁。可以在加上「*[SIGNA-TURE PAGE FOLLOWS]*」以說明下一頁還有簽名欄。另外簽名頁的頁尾也可以註記「Signature page to XXX Agreement dated, 20xx」。較能一目瞭然到底簽的是什麼契約。

　　另外，也可在每頁頁緣處讓被授權人簽上姓名的英文字母縮寫，以確保契約內容不會被偷天換日。簽上姓名縮寫也可以用來證明當事人確實仔細讀過契約，例如在美國要放棄聲請陪審團審判的權利，簽上姓名縮寫就表示是自願、有意的放棄^{（註15）}。

　　以下提供例句：

例句十一：

　　To evidence the parties' agreement to this Agreement, they have signed and delivered it on the date set forth in the preamble^{（註16）}.

OWNER

_____

John Doe

LICENSEE
ABC LLC

15　Stark 前揭著第 195-196 頁。
16　Stark 前揭著第 449 頁。

By: DEF LLC, its manager
By its duly authorized signatory

_____

Name: Jim Cheng
Title: President

[Signature page to License Agreement dated   ,20xx]

2. 附件（Attachment）

　　中文的「附件」在英文中有很多講法，例如「appendix」、an-nex」、「attachment」、「annexure」。附件的類型也可以再區別為「exhibit」跟「schedule」。但對於兩者的定義卻有不同的說法，有主張「exhibit」是本來就獨立存在的文件，例如公司組織圖、章程等，只是在這份契約裡面拿來用，而「schedule」的內容本來是契約裡的文句，但把它移到契約後面作為附件（註17）。但也有人主張「schedule」只限於聲明與擔保事項，其他的附件則稱為「exhibit」（註18）。所以即使美國人對這個區別也並不一致，我們並不需要區分「exhibit」跟「schedule」，直接稱為「appendix」或「attachment」即可。

　　附件有便利與保密兩層功能。太瑣細的資訊，可以移到契約之後作為附件，以增加契約的流暢與易讀性。有機密性的文件，如果放到契約本文裡面，傳送給所有相關的人，可能有所顧忌，列為附件，限制讀取

17　Adams 前揭著第 105-110 頁。
18　Stark 前揭著第 282-285 頁。

權限就可以解決這個問題，例如法規可能要求必須把重要契約文件傳一份給主管機關（例如股票上市公司要傳給證券機關），但主管機關許可將敏感的文件列為附件，而不必傳送附件。

英文契約中附件記得要編號，用「A、B、C」或「1、2、3」都可以。

攸關權利義務的重要事項，或者法律要求約定要寫得很醒目（例如放棄聲請交陪審團審判的權利，或者消費者契約中的特約事項）避免放到附件中，以免日後引起紛爭。附件不能寫某方應做什麼，或某方有什麼權利，不要出現「shall」、「may」這些詞。但是可以補充契約本文裡面權利或義務的具體內容。

附件最好也加上一些防偽機制，例如要求簽姓名縮寫，以避免被掉包或爭執是偷偷加上去的。

有時候附件的內容可能是浮動的，所以甚至不會貼在契約後面，例如特定行庫定期公告的基準利率，這類附件被稱為「虛擬的」（virtual）。參照到這種附件時，最好費些筆墨寫清楚，特定虛擬附件的內容是契約的構成部分，雙方確認都受其拘束（「The content of XXX is part of this Agreement and binds each Party.」）。不要只說「subject to」或該項文件可以在哪個網址查得到。尤其如果虛擬附件是當事人一方作成的，他方事後可能用很多理由爭執。如果這類虛擬附件是當事人一方作成的，每次變更最好還是通知其他當事人。如果虛擬附件是第三人作的，在契約寫清楚，各方當事人自己隨時查詢，互不負通知義務，自然最為妥當。

3. 契約修訂（Amendment）

簽約之後，有時候也會遇到需要修訂的情況。在英文契約中，修訂稱為「amendments」。

(一) 範例

修訂契約就像是寫一份較為精簡的新契約，雖然精簡，但也必須有模有樣。以下提供範例：

> ### 例句十二： （註19）
>
> **AMENDMENT No.1 TO EMPLOYMENT AGREEMENT**
> **First Amendment**, dated February 10, 2018, to the Employment Agreement , dated April 5, 2015, between John Doe, domiciled at 1602 Roosevelt Rd., Taipei, Taiwan (the "Employee"), and ABC LLC, a Taiwan limited liability company (the "Company").
> **Background**
> The Employee has served in the Company as Manager of Operation Department for about 3 years.
> The Company now desires to promote the Employee as General Manager.
> The parties desire to amend the Employment Agreement dated April 5, 2015, as amended (the "Existing Agreement"), to

---

19 本範例引用 Stark 前揭著第 362-367 頁及 Adams 前揭著第 375-380 頁範例文句，對照辨正並改寫。

reflect the Employee's new duties, compensation, and other matters.

Accordingly, the parties agree as follows:

1. Definitions. Capitalized terms used in this Amendment without definitions have the meaning assigned to them in the Existing Agreement.

2. Amendment to Section 3.5. Section 3.5of the Existing Agreement is amended by deleting that Section and inserting in its place the following:

3.5 Term. The term of employment under this Agreement is from February 15, 2018 through February 14, 2023 (the "Employment Term")

(The employment term of the existing agreement was from April 15, 2015through April 14, 2020.)

3. Amendment to Section 3.6. Section 3.6 of the Existing Agreement is amended to read as follows:

3.6 Salary. The Company shall pay the Employee at a rate of 10,000 a month during the Employment Term. The Company shall pay the monthly salary on the last business day of each month.

(Salary amount is modified.)

4. Amendment to Section 5. Section 5 of the Existing Agreement is amended by deleting original words and inserting in its place "Section 5 has been intentionally omitted."

5. Addition of Section 7. The Existing Agreement is amended by inserting the following provision as Section 7 after the end of the Existing Agreement's Section 6:

7. Location of General Manager's Office. The Company shall not locate General Manager's office outside of the greater Metropolitan Taipei region without the Employee's consent, which consent the Employee shall not unreasonably withhold.

6. Renumbering of Sections. The Existing Agreement is amended to change the section numbers of Sections 8, 9, 10, and 11 to Sections 9, 10, 11, and 12, respectively.

7.Authorization. The Company represents that its board of directors has authorized the amendments.

8. Original Representations and Warranties. The Employee represents and warrants that the original representations and warranties made by him in the Existing Agreement are also true as of the signing of this amending agreement.

9. Amendments' Effectiveness. The amendments in this amending agreement are effective upon the satisfaction of the following condition: The Employee must have delivered his EMBA diploma to the Company. If this condition is not satisfied before February 15, 2018, this amending agreement terminates, and the Existing Agreement continues unchanged and in full force.

10. Continued Effectiveness of the Existing Agreement. Except as amended by this amending agreement, the Existing Agreement continues unchanged and in full force.

EMPLOYEE

_____

John Doe

COMPANY

ABC LLC

By: DEF LLC, its manager

By its duly authorized signatory

_____

Name: Jim Cheng

Title: President

[Signature page to License Agreement dated   ,20xx]

(二) 說明

　　契約修訂跟寫一份新的契約一樣，一開始都必須交代當事人，且宜說明修訂背景。並帶出雙方合意的文字。

　　修訂的契約同樣有定義條款，但可以完全參照原契約。

　　至於修訂的主要內容，有兩種很常見，但是並不恰當的修訂方法。第一種是講前一份契約第幾條第幾項的什麼字改成什麼字、加上什麼字、刪除什麼字。顯然這種修訂方法沒辦法讓人一目瞭然，好像做猜謎遊戲一樣，萬一猜謎本身寫錯，那就更麻煩了。第二種方法是只講契約修訂的結果變成什麼樣子，但卻沒有交代哪幾條被修改。

　　比較好的方法是說明清楚哪幾條被修訂，而且修訂後完整的條款為何，當然附上舊條款作對照是更完善。刪除舊條款可以寫「原條文特意刪除。」，就不需要更動條號，但新增條款則必須處理條次變動的問

題。

　　除了修訂內容外，也可能重中原契約的聲明擔保仍屬有效、或加上修訂內容的生效條件，並說明原契約未更動的部份仍屬有效。

　　最後一樣是當事人簽名。

附件

附件一　中英對照保密協議

Non-Disclosure Agreement

保密協議

This Non-Disclosure Agreement (hereinafter, this "Agreement") is made on [MM/DD/YYYY] (hereinafter, the "Effective Date") by [company name], a [company form] incorporated under laws of [jurisdiction], with [company code of xxxxx], and registered office at [address] (hereinafter, the "Discloser"), and [company name], a [company form] incorporated under laws of [jurisdiction], with [company code of xxxxx], and registered office at [address] (hereinafter, the "Recipient"). The Discloser and Recipient are referred to hereafter as a "Party," individually, and the "Parties," collectively.

本保密協議（以下簡稱「本協議」）於〔年/月/日〕（以下簡稱「生效日」）簽訂，當事人為〔公司名稱〕，係依據〔法域〕設立的〔公司類型〕，〔公司編碼為xxxx〕，註冊地址位於〔地址〕（以下簡稱「揭露方」），以及公司名稱〕，係依據〔法域〕設立的〔公司類型〕，〔公司編碼為xxxx〕，註冊地址位於〔地址〕（以下簡稱「收受方」）。揭露方與收受方已下個別稱為「一方當事人」，合稱為「各方當事人」。

Whereas, the Parties wish to [conduct proposed transactions] (hereinafter, the "Purposes"),

Whereas, for the Purposes, the Discloser may desire to disclose or make available certain information and materials of confidential and/ or proprietary nature to the Recipient or its Representatives (as defined below) in connection with certain discussions, negotiations or dealings between the Parties relating to the Purposes,

Whereas, the Parties agree to protect and preserve the confidential and/or proprietary nature of the Discloser's Proprietary Information (as defined below) that may be disclosed or made available to the Recipient .

In consideration of the foregoing and the rights and obligations set forth herein, the Parties hereby agree as the following:

鑑於，雙方當事人有意〔進行預期交易〕（以下簡稱「目的事項」），

鑑於，為目的事項，揭露方可能有意向收受方或其代表人（如下定義）揭露或提供特定屬機密性質或私有性質，且有關於各方當事人就目的事項之討論、協商或交易的資訊與材料，

鑑於，雙方當事人合意保護並保存可能向收受方揭露或提供的揭露方私有資訊（如下定義）。

以前述事項與本協議之權利及義務為約因，各方當事人合意如下：

1. Definitions

1.1　"Person" shall be broadly interpreted to include, without limitation, any corporation, company, partnership, other entity or individual.

1.2　"Proprietary Information"

(a) Positive Definitions. Proprietary Information means any and all information of the Discloser that is not known to Persons generally involved in the information of this type, irrespective of whether or not such information (1) can be used in the course of production, sales, or operations, (2) has any economic value, actual or potential, due to its secretive nature, and (3) is disclosed in tangible or intangible form. In addition, Proprietary Information also includes (1) the fact that discussions or negotiations are taking place concerning the Purposes, (2) the proposed terms and conditions of the Purposes (including any financial terms and conditions) and the status thereof, and (3) the existence, context, and scope of this Agreement.

(b) Exclusion. Proprietary Information does not include information that: (1) is or becomes generally available to the public other than as a result of any disclosure or other action or inaction by the Recipient in breach of this Agreement (including any disclosure or other action or inaction by Representatives of the Recipient that would constitute a breach of this Agreement if undertaken by the Recipient itself); (2) is or becomes known or available to the Recipient or any of its Representatives on a non-confidential basis from a source (other than the Discloser or any of its subsidiaries, affiliates or Representatives) that, to the best of the knowledge of the Recipient, is not prohibited from disclosing such Proprietary Information to the Recipient by a contractual, legal or fiduciary obligation; or (3) is or was independently developed by the Recipient or any of its Representatives without violation of any obligation under this Agreement.

1.3　"Representatives" means as to any Person, its directors, officers, employees, agents and advisors (including, without limitation, financial advisors, financing sources, attorneys, accountants and any one which is authorized to act on behalf of the Persons).

1. 定義。

1.1　「人」應廣義地解釋為包括，但不限於任何企業、公司、合夥、其他實體或個人。

1.2　「私有資訊」

(a) 正面定義. 私有資訊指揭露方任何非一般涉及此類資訊之人所知的任何一切資訊，無論此等資訊是否(1)可用於製造、銷售或營運，(2)因其敏感性質而具有任何實際或潛在的經濟價值，以及(3)以有形方式或無形方式揭露。此外，私有資訊也可能包括(1)正發生有關目的事項的討論或協商之事實，(2)有關目的事項而提出的條款與條件（包括任何財務條款與條件），及其進度，以及(3)本協議之存在、背景與範圍。

(b) 排除資訊. 私有資訊並不包括以下資訊：(1)現在或將來非因收受方違反本協議之揭露或其他作為或不作為而變為一般大眾可得之資訊（包括代表人之揭露或其他作為或不作為，而若由收受方自行揭露或作為或不作為即構成違約者）；(2)收受方或任何其代表人現在或將來以非保密的基礎由（揭露方

或其任何子公司、關係企業或代表人以外的）某來源獲知或取得，而且就收受方所知，該來源並未因為契約、法律或忠實義務而不得揭露該等私有資訊；或(3)現在或過去由收受方或其任何代表人於未違反本協議之義務下獨立開發。

1.3　「代表人」指任何人、其董事、主管、員工、代理人與顧問（包括但不限於財務顧問、融資管道、律師、會計師，及其他任何獲授權代表該等人士行為之人）。

2. NON-DISCLOSURE AND LIMITED USE.

2.1　Non-Disclosure. Without prior written consent of the Discloser and except as otherwise required by applicable law, the Recipient shall keep, and shall cause its Representatives to keep, all Proprietary Information confidential and shall not disclose or reveal, and shall cause its Representatives not to disclose or reveal, in any manner whatsoever, in whole or in part, any Proprietary Information to any Person, other than to its Representatives who are actively and directly participating in its evaluation of the Purposes or who otherwise need to know the Proprietary Information for assessment and evaluation relevant with the Purposes and who are bound by restrictions regarding the disclosure and use of such Proprietary Information (either contractual, legal or fiduciary) owed to the Discloser, the Recipient or any their respective Representatives that are comparable to and no less restrictive than those set forth in this Agreement. The Recipient shall inform all of its Representatives and shall cause its Representatives to inform their Representatives who receive Proprietary Information hereunder of the confidential nature of such information and the Purposes, as well as the terms of this Agreement. The Recipient shall not, and shall cause its Representatives to not, use any Proprietary Information for any purpose other than to evaluate the Purposes or in connection with the consummation of the Purposes. Te Recipient shall be responsible for any breach of the terms of this Agreement by it or its Representatives.

2.2 Degree of Care. The Recipient shall take the same degree of care that it uses to protect its own confidential and proprietary information of similar nature and importance (but in no event less than reasonable care) to protect the confidentiality and avoid the use, disclosure, publication, or dissemination of the Proprietary Information of the Discloser. The Recipient shall not, and shall cause its Representatives not to decompile, disassemble, or otherwise reverse engineer (except to the extent expressly permitted by applicable law, notwithstanding a contractual obligation to the contrary) any Proprietary Information or any portion thereof, or determine or attempt to determine any source code, algorithms, methods or techniques embodied in any Proprietary Information or any portion thereof. The Recipient shall not use Proprietary Information for any purpose or in any manner that would constitute a violation of any laws or regulations, including without limitation the export control laws of the United States.

2.3 Designated Representatives. Neither the Recipient nor its Representatives shall initiate or maintain contact with any officer, director, stockholder, employee or agent of the Discloser or its subsidiaries regarding the Purposes, except with the express consent of the Discloser. All (i) communications regarding the Purposes, (ii) requests for additional information, (iii) requests for on-site access or management meetings and (iv) discussions or questions regarding procedures, will be submitted or directed to the Representatives designated by the Discloser.

2.4　Compelled Disclosure of Proprietary Information. If the Recipient or any of its Representatives are requested pursuant to, or required by, applicable law or regulation (including, without limitation, any rule, regulation or policy statement of any national securities exchange, market or automated quotation system on which any of the Recipient's securities are listed or quoted) or by legal process to disclose any Proprietary Information, or any other information concerning the Discloser, its subsidiaries or affiliates, the Recipient shall provide the Discloser with prompt notice of such request or requirement, in order to enable the Discloser (a) to seek an appropriate protective order or other remedy, (b) to consult with the Recipient with respect to the Discloser's taking steps to resist or narrow the scope of such request or legal process or (c) to waive compliance, in whole or in part, with the terms of this Agreement. In the event that such protective order or other remedy is not obtained, or the Discloser waives compliance, in whole or in part, with the terms of this Agreement, the Recipient or its Representatives, as the case may be, shall use commercially reasonable efforts to disclose only that portion of the Proprietary Information which the Recipient is advised by legal counsel is legally required to be disclosed and exercise its commercially reasonable efforts to obtain reliable assurances that confidential treatment will be accorded to the Proprietary Information so disclosed.

2.5　Attorney-Client Privilege. To the extent that any Proprietary Information may include material subject to the attorney-client privilege, work product doctrine or any other applicable privilege concerning pending or threatened legal proceedings or governmental investigations, the Parties understand and agree that they have a commonality of interest with respect to such matters and it is their desire, intention and mutual understanding that the disclosure of such material is not intended to, and shall not, waive or diminish in any way the confidentiality of such material or its continued protection under the attorney-client privilege, work product doctrine or other applicable privilege and any such Proprietary Information shall remain entitled to all protection under these privileges, this Agreement, and under the joint defense doctrine. Nothing in this Agreement obligates any Party to reveal material subject to the attorney-client privilege, work product doctrine or any other applicable privilege.

2.6　Definitive Agreement Until a definitive agreement regarding the Purposes has been executed by the Parties hereto, neither Party hereto shall be under any legal obligation or have any liability to the other Party of any nature whatsoever with respect to the Purposes by virtue of this Agreement or otherwise (other than with respect to the confidentiality and other matters set forth herein). Each Party hereto and its Representatives (i) may conduct the process that may or may not result in the Purposes in such manner as such Party, in its sole discretion, may determine (including, without limitation, negotiating and entering into a definitive agreement with any third party without notice to the other Party) and (ii) reserves the right to change (in its sole discretion, at any time and without notice to the other Party) the procedures relating to the Parties' consideration of the Purposes (including, without limitation, terminating all further discussions with the other Party). In this Agreement, the term "definitive agreement" does not include an executed letter of intent or any other preliminary written agreement in principle.

2.7 No Representations or Warranties Regarding Proprietary Information. Subject to the terms and conditions of a definitive agreement regarding the Purposes and without prejudice thereto, each Party acknowledges that neither the other Party nor its Representatives nor any of the officers, directors, employees, agents or controlling Persons of such Representatives makes any express or implied representation or warranty as to the completeness of the Proprietary Information or any use thereof. Each Party hereby expressly disclaims all such warranties, including any implied warranties of merchantability and fitness for a particular purpose, non-infringement and accuracy, and any warranties arising out of course of performance, course of dealing or usage of trade. The Recipient shall not be entitled to rely on the completeness of any

Proprietary Information but shall be entitled to rely solely on such representations and warranties regarding the completeness of the Proprietary Information as may be made to it in a definitive agreement relating to the Purposes, subject to the terms and conditions of any such agreement, should the discussions between the Parties progress to such a point.

2. 禁止揭露與使用限制。

　　2.1　禁止揭露。未經揭露方事先書面同意，或非屬相關法律要求，收受方應自行，並應督促其代表人對各種私有資訊予以保密，且無論以任何方式均不得自行揭露或洩露，並應督促其代表人不得揭露或洩露任何私有資訊之一部或全部予任何人，但收受方之代表人，且積極並直接參予其目的事項評估工作者，或因其他原因需要知悉私有資訊以進行有關目的事項的衡量與評估，且對揭露方、收受方或其個別之代表人負有關於此等私有資訊揭露與使用之限制（無論契約、法律或忠實義務上的限制），且其受限程度不低於本協議者不在此限。收受方應告知收到私有資訊的代表人並督促該等代表人該等資訊的機密性質以及目的事項，以及本協議之條款。除為了評估目的事項，或有關目的事項之施行外，無論任何目的，收受方均不得自行，亦應督促其代表人不得使用任何私有資訊。收受方對於自己或其代表人違反本協議條款之行為應負擔責任。

　　2.2　注意程度。收受方應採取與其用以保護自有性質與重要程度相近的機密與私有資訊相同程度之注意（但無論如何不得低於合理程度的注意）以保護揭露方的私有資訊之機密性，並避免其使用、揭露、公開或散布。收受方不得、亦應督促其代表人不得對任何私有資訊或其任何部分進行反編譯、拆解、或以其他方法進行逆向工程（但相關法律明

確許可者不在此限，縱契約課予相反之義務亦同）或對任何私有資訊所中所含有的任何來源碼、公式、方法或技術，或其部分加以判斷或試圖判斷。收受方不得為可能違反任何法律或法規之目的或以可能違反任何法律或法規之方式使用私有資訊，包括違反美國出口管制法律之情況。

2.3　指定代表人。除經揭露方明確書面同意之外，收受方或其代表人均不得針對有關目的事項對揭露方或其子公司之任何主管、董事、股東、員工或代理人主動發起或維持聯繫。所有(i)關於目的事項之通訊 (ii)要求提他資訊(iii)要求現場訪查或管理會議，以及(iv)關於程序的討論或問題，應向揭露方指定的代表人提出。

2.4　強制揭露私有資訊。如果收受方或其任何代表人依據相關法律法規（包括但不限於有關收受方證券上市或報價交易所依循的任何國家證券交易、市場或自動報價系統任何規則、法規或政策聲明）或受其強制，或因為法律程序而必須揭露任何私有資訊，或有關於揭露方、其子公司或關係企業的任何其他資訊，收受方應立即通知揭露方有關此等請求或要求，以使揭露方(a)得聲請適當的保護令或尋求其他救濟，(b)徵詢收受方有關揭露方為避免此種要求或法律程序或限縮其範圍，可採取之步驟，或(c)免除遵循本協議之部分或全部條款。如果無法獲得此等保護令或其他救濟，或者揭露方免除遵守本協議條款之義務，收受方或其代表人視情況應盡商業上合理努力僅揭露收受方依其法律顧問之建議認定在法律上有義務揭露之部分，並盡商業上合理努力以獲得可靠之確保使揭露之資訊可獲得保密處理。

2.5　律師與當事人間的保密特權.在私有資訊可能含有受律師與當事人間保密特權、工作成果法則、或有關繫屬中或擬提出法律程序或政府調查而適用的其他任何特權保護之材料的範圍內，各方當事人瞭解並同意對於此等事項具有共同的利益，而且渠等均希望、有意並相互瞭解

此種材料不宜揭露，而且不以任何方式免除或降低此等材料的機密程度，或其依據律師與當事人間保密特權、工作成果法則或其他可適用的特權而受保護，而任何此等私有資訊應持續有權受到此等特權、本協議與共同抗辯特權之保護。本協議條款無任何內容使任何當事人有義務揭露受律師當事人間保密特權、工作成果法則或其他任何可適用之特權保護的材料。

2.6　確定契約. 在本協議當事人就目的事項簽訂確定契約之前，本協議之當事人不因本協議或其他事由而對他方當事人就目的事項負擔任何性質的法律上義務或責任（但本協議之保密與其他事項不在此限）。本協議各方當事人與其代表人(i)得以該當事人全權裁量之決定進行或能或不能達成目的事項之過程（包括但不限於不經通知他方當事人逕與第三人協商並簽訂確定契約），並(ii)保有權利得（依其全權裁量，隨時不經通知他方當事人）變更有關當事人審酌目的事項之程序（包括，但不限於終止與他方當事人間的後續討論）。在本協議，「確定契約」並不包括簽署意向書或其他任何初步的書面合意。

2.7　關於私有資訊無任何聲明或擔保。Subject to the terms and conditions of a definitive agreement regarding the Purposes and without prejudice thereto, each Party acknowledges that neither the other Party nor its Representatives nor any of the officers, directors, employees, agents or controlling Persons of such Representatives makes any express or implied representation or warranty as to the completeness of the Proprietary Information or any use thereof. 在關於目的事項的確定契約之條款與條件限制下且在不妨礙確定契約之效力下，各方當事人確認他方當事人或其代表人、或其任何主管、董事、員工或改等代表人之控制人士對於私有資訊之完整性及其使用並無任何明

示或默示的聲明或擔保。各方當事人特此明確排除任何此等擔保，包括任何適銷性與適於特定目的、不侵權與準確性之默示擔保，以及出於履約過程、交易往例或行業習慣的任何擔保。收受方無權信賴任何私有資訊的完整性，僅有權信賴確定契約中可能包含關於私有資訊完整性的完整性之聲明與擔保，但應受該等契約之條款與條件之限制，若各方當事人間的討論進展到簽署確定契約的程度。

3. NO SOLICITATION.

Neither the Recipient nor its affiliates may at any time until the one year anniversary of the Effective Date, directly or indirectly, employ or solicit for employment (i) any key technical or management personnel of the Discloser that has first been introduced by the Discloser to the Recipient in connection with the Purposes or who was otherwise substantively involved in the discussions of the Purposes or (ii) any other Person who is now employed as an officer of the Discloser or any of its affiliates; provided, that the foregoing restriction shall not prohibit the Recipient or its Representatives from making general public solicitations for employment for any position or from employing any employee of the Discloser who either responds to such a general solicitation for employment or otherwise contacts the Recipient on his or her own initiative and without solicitation by the Recipient in contravention of the above restriction.

3. 禁止招攬

收受方及其代表人在生效日後一週年內均不得直接或間接聘僱或招攬下列人員 (i) 先前由揭露方首次向收受方為有關目的事項介紹，或曾經

以其他方式實質涉入目的事項之討論的揭露方任何關鍵技術人員或管理人員，或(ii)任何現在受揭露方或其關係企業聘僱擔任主管之人士；但前述限制不應限制收受方或其代表人為任何職位進行公開招募，亦不限制聘僱因應一般招募之揭露方員工或未經收受方違反前述限制的招攬行為而自行主動聯繫收受方的揭露方員工。

4. No Public Disclosure

4.1　No Public Disclosure Required. Each Party expressly confirms and agrees that, as of the date hereof, it is not required to make any public disclosure with respect to (a) the Purposes (or the terms or conditions or any other facts relating thereto), (b) any item of Proprietary Information (or the fact that Proprietary Information has been made available to such Party), or (c) any discussions or negotiations taking place between the Parties with respect to the Purposes, pursuant to any applicable laws and regulations. If, after the date of this Agreement, either Party determines that any such disclosure is required, no such disclosure shall be made unless and until such Party consults with the other Party regarding the necessity and form of any such disclosure, and provides the other Party a reasonable opportunity to review the proposed disclosure and comment thereon.

4.2　Restrictions on Sales of Securities. Each Party is aware, and will advise its Representatives who are informed of the matters that are the subject of this Agreement, of the restrictions imposed by applicable securities laws on the purchase or sale of securities by any Person who has received material, non-public information from the issuer of such securities and on the communication of such information to any other Person when it is reasonably foreseeable that such other Person is likely to purchase or sell such securities in reliance upon such information. However, this will not prevent the Recipient from purchasing stock through its benefit plans in the ordinary course of business.

4. 無須公開揭露

4.1　無須公開揭露。各方當事人明示確認並合意，在本協議簽訂日，依據任何相關法律與規則無須公開揭露(a)目的事項（或其相關條件或其他事實），(b)私有資訊之任何項目（或已向該方當事人提供私有資訊之事實），或(c)全體當事人間發生有關目的事項之討論或協商。如果，在本協議簽訂之日後，任一方判斷有此等揭露之義務，該方當事人應就此等揭露之必要性與方式徵詢他方當事人意見，並提供他方當事人合理機會以審核擬行的揭露並表示意見之後，始得揭露。

4.2　買賣有價證券之限制。各方當事人知悉，對於明瞭本協議主旨事項之代表人並將告知相關證券法律對於任何已收到證券發行人重大非公開資訊之此等人士課予有關買賣有價證券之限制，以及有關將此等資訊傳達給其他任何可合理期待將倚賴此等資訊進行有價證券買賣的其他任何人士。然而本項規定並不禁止收受方透過自訂的受益計畫依常規交易購買股票。

5. OWNERSHIP.

All Proprietary Information (including, without limitation, all copies, extracts and portions thereof) is and shall remain the sole property of the Discloser. The Recipient does not acquire (by license or otherwise, whether express or implied) any intellectual property rights or other rights under this Agreement or any disclosure hereunder, except the limited right to use such Proprietary Information in accordance with the express provisions of this Agreement. All rights relating to the Proprietary Information that are not expressly granted hereunder to the Recipient are reserved and retained by the Discloser.

5. 所有權。

　　所有私有資訊（包括但不限於所有影本、節本與其部分）均始終屬於揭露方專有之財產。除了依據本協議明示條款使用私有資訊之有限權利外，收受方並不因為本協議，或依據本協議而進行的任何揭露而取得（無論依授權或其他方式，無論明示或默示）任何智慧財產權，或其他權利。並未依據本協議明確授予的其他各種有關私有資訊之權利均由揭露方保留並保有。

6. TERM.

　　Except as otherwise provided herein, the obligations of this Agreement, including the restrictions on disclosure and use, shall expire on the second anniversary of the Effective Date; provided that Sections 2.5, 2.6, and 2.7 and Articles 4, 5, 6, 7, 8 and 9 shall survive any expiration of this Agreement.

6. 效期。

　　除本協議另有約定外，本協議之義務，包括揭露與使用之限制，於生效日後滿第二年之日到期；但第2.5,2.6,2.7條與第4,5,6,7,8,及第9條在本協議到期後繼續有效。

7. REMEDIES.

The Recipient agrees that, due to the unique nature of the Proprietary Information, the unauthorized disclosure or use of the Proprietary Information may cause injury to the Discloser, the extent of which will be difficult to ascertain and for which there may be no adequate remedy at law. Accordingly, the Recipient agrees that the Discloser, in addition to any other available remedies, may have the right to seek an immediate injunction and other equitable relief enjoining any breach or threatened breach of this Agreement as ordered by the court. The Recipient shall notify the Discloser immediately if Recipient has reason to believe that any Person who has had access to the Proprietary Information (including the Recipient or any of its Representatives) has violated or intends to violate the terms of this Agreement or otherwise disclose any Proprietary Information in violation of the terms hereof. Any and all remedies herein expressly conferred upon a Party will be deemed cumulative with and not exclusive of any other remedy conferred hereby, or by law or equity upon such Party, and the exercise by a Party of any one remedy will not preclude the exercise of any other remedy.

7. 救濟。

收受方同意,因私有資訊之性質,私有資訊若未經同意而揭露或使用,可能對揭露方造成損害,而且損害程度可能難以確認,因此在法律上難有適當的救濟。因此,收受方同意揭露方除了其他可請求的救濟之外,有權聲請即時的強制處分與其他衡平救濟由法院下令禁止任何違反本協議或有違約之虞的行為。收受方如果有理由相信任何可取得私有資

訊之人士（包括收受方或其代表人）已違反本協議之條款或以其他方式違反本協議之條款而揭露任何私有資訊，收受方應立即通知揭露方。本協議明確許可一方當事人行使之救濟均屬累積性的，而不排除本協議、法律或衡平規則許可該當事人行使的其他救濟，而當事人行使某項救濟並不妨礙行使其他任何救濟的權利。

8. RETURN OF MATERIALS.

If either Party hereto shall determine that it does not wish to proceed with the Purposes, such Party shall promptly advise the other party of that decision. In that case, or if the Purposes otherwise is not consummated for any reason, the Recipient shall, upon the Discloser's written request, promptly deliver to the Discloser all Proprietary Information, and, at the Discloser's sole election, return or destroy (provided that any such destruction shall be certified by a duly authorized Representative of the Recipient) all copies, reproductions, summaries, analyses or extracts thereof or based thereon (whether in hard-copy form or on intangible media, such as electronic mail or computer files) in the Recipient's or any of its Representatives' possession; provided, that if a legal proceeding has been instituted to seek disclosure of the Proprietary Information, such material shall not be destroyed until the proceeding is settled or a final judgment with respect thereto has been rendered. However, the Recipient may retain in the office of its legal counsel, one copy of Proprietary Information for record only. Notwithstanding the return or destruction of any Proprietary Information, or documents or material containing or reflecting any Proprietary Information, the Parties will continue to be bound by their obligations of confidentiality and other obligations hereunder for the term of this Agreement (or such other term as may be applicable to the specific obligation), except as otherwise specifically provided herein.

8. 返還材料。

如果本協議任一方當事人決定無意續行目的事項，該方當事人應立即告知他方當事人其此項決定。在該情況，或者如果因任何原因未實行該等目的事項，在揭露方書面要求時，收受方應立即將所有私有資訊交還揭露方，並依揭露方之全權裁量返還或銷毀（但銷毀時應由正式授權的收受方代表人證明之）收受方或其代表人持有的一切影本、複製品、摘要、分析或摘錄本，或依私有資訊製作者（無論是否為實體影印或無形媒體資料，例如電子郵件或電腦檔案）；但如果已發起法律程序聲請揭露私有資訊，在程序達成和解或作出最終裁判之前不得銷毀此等材料。然而收受人得在其法律顧問的辦公室專為存檔之目的保留一份私有資訊 。縱使返還或銷毀私有資訊、或包含或反映任何私有資訊內容的文書或材料，除非本協議另有明訂，各方當事人仍將繼續依據本協議之條款受本協議所定保密義務與其他義之約束（或有關於特定義務應適用之條款）。

9. MISCELLANEOUS.

9.1 Entire Agreement. This Agreement constitutes the entire agreement between the Parties concerning the confidentiality of the Proprietary Information in connection with the Purposes and related matters and supersedes all prior or contemporaneous representations, discussions, proposals, negotiations, conditions, communications and agreements, whether oral or written, between the Parties relating to the same and all past courses of dealing or industry custom.

9.2　Beneficiaries. This Agreement shall inure to the benefit of and be binding upon the Recipient and the Discloser and their respective successors and permitted assigns.

9.3　Amendments and Waivers. No amendment, modification or waiver of any provision of this Agreement shall be effective unless in writing and signed by duly authorized signatories of the Parties. The waiver by either Party of a breach of or a default under any provision of this Agreement shall not be construed as a waiver of any subsequent breach of or default under the same or any other provision of this Agreement, nor shall any delay or omission on the part of either Party to exercise or avail itself of any right, power, privilege or remedy that it has or may have hereunder operate as a waiver thereof, nor shall any single or partial exercise thereof preclude any other further exercise of any such right, power, privilege or remedy hereunder.

9.4　Choice of Law. This Agreement shall be governed by and construed in accordance with the laws of Singapore without regard to conflicts of laws principles, and except for laws of Singapore, all state, national, regional, international laws, regulations, treaties, conventions shall be excluded.

9.5　Jurisdiction. Any dispute arising out of or in connection with this Agreement, including any question regarding its existence, validity or termination, shall be referred to and finally resolved by arbitration administered by the Singapore International Arbitration

Centre ("SIAC") in accordance with the Arbitration Rules of the Singapore International Arbitration Centre ("SIAC Rules") for the time being in force, which rules are deemed to be incorporated by reference in this clause. The seat of the arbitration shall be Singapore. The Tribunal shall consist of one (1) arbitrator. The language of the arbitration shall be English.

9.6 Severability. In the event that any of the provisions of this Agreement shall be held by a court or other tribunal of competent jurisdiction to be invalid or unenforceable, the remaining portions hereof shall remain in full force and effect and such provision shall be enforced to the maximum extent possible so as to effect the intent of the Parties, and shall in no way be affected, impaired or invalidated.

9.7 Notices. Any notice or other communication required or permitted to be delivered under this Agreement shall be in writing and shall be deemed effectively given: (i) upon Personal delivery to the Party to be notified; (ii) when sent by confirmed telex or facsimile if sent during normal business hours of the Recipient, if not, then on the next business day; (iii) three (3) days after having been sent by registered or certified mail, return receipt requested, postage prepaid; or (iv) one (1) day after deposit with a nationally recognized overnight courier, specifying next day delivery.

9.8 Counterparts. This Agreement may be executed in one or more counterparts, each of which will be deemed to be an original copy of this Agreement and all of which, when taken together, will be deemed to constitute one and the same instrument.

9. 附則。

9.1　完整合意。本協議構成各方當事人間目的事項相關私有資訊保密與關連事項之完整合意，並取代當事人之間就相同事項先前或同時的各種聲明、討論、提議、協商、條件、通訊與合意，無論是口頭的或書面的，也取代各種先前交易往例或產業習慣。

9.2　受益人。本協議應為收受方與揭露方及其個別繼受人與獲許可之受讓人的利益繼續存在，並對渠等有拘束力。

9.3　增補與棄權。本協議任何條款之增補、修訂或棄權非以書面由各方當事人正式授權簽約人書面簽名者不生效力。任一方當事人放棄追究違反或違背本協議任一條款者，不得解釋為對本協議相同條款或任何其他條款嗣後的違反行為予以追究之權利，任一當事人遲延或疏未行使或利用任何權利、權能、特權或主張其依據本協議享有之救濟者，不得視為棄權，而僅行使某項權利或部分行使權利並不排除後續任何行使權利、特權或主張救濟之權能。

9.4　法律選擇。本協議應受新加坡法律規範並依該法律解釋，但不適用其法律衝突原則，且除新加坡法律之外，其他所有州、國、區域或國際法律、法規、條約、公約均應排除適用。

9.5　管轄。任何起因於或有關於本協議之糾紛，包括任何有關其存在、效力、或終止者，均應提交新加坡國際仲裁中心（「SIAC」）依據當時有效的新加坡國際仲裁中心仲裁規則（「SIAC規則」）進行仲裁並最終解決，該規則應視同已納入本條。仲裁地應為新加坡。仲裁庭應由一(1)位仲裁人組成。仲裁語言應為英文。

9.6　效力可分。如果本協議有任何條款經法院或其他有管轄權之裁判機關認定屬於無效或者不能強制履行，本協議其餘條款應繼續完全

有效，且此等條款應在可能的最大程度內予以執行以使當事人的意向得以落實，而且無論如何均不受影響、妨礙或變為無效。

9.7　通知。依據本協議應發送或得發送的任何通知或其他通信均應以書面為之，且在下列時點視為有效發出：(1)在專人送達給應受通知之一方當事人時；(ii)如果在收受方通常營業時間發送時以確認的電傳或傳真時為準，如果不是在收受方的通常營業時間發送，則在次一營業日，(iii)如果以掛號或任證，並附回執且預付郵資方式發出，在寄出後的三日；或者(iv)如果以本國認可的隔夜快遞發送載明次日送達，則在交寄後一(1)日。

9.8　原本。本協議得以一份或多份簽署，各分均視為本協議之原本，且共同構成單一而相同之文書。

[SIGNATURES FOLLOW ON THE NEXT PAGE]

[簽名頁在次頁]

IN WITNESS WHEREOF, the Parties have duly authorized and caused this Non-Disclosure Agreement to be executed as follows:

鑒於上述事項，各方當事人茲授權並促使以下人適簽署本保密協議：

The Discloser

揭露方

[Company Name]

[公司名稱]

By_____

簽約人

　　[Signature]

　　〔簽名〕

Name_____

姓名

　　[Print]

　　[正楷]

Title

職銜

The Recipient

收受方

[Company Name]

[公司名稱]

By_____

簽約人

   [Signature]

   〔簽名〕

Name_____

姓名

   [Print]

   [正楷]

Title

職銜

附件二　中英對照經銷協議

Distribution Agreement

經銷協議

This Distribution Agreement (hereinafter, this "Agreement") is made as of this [YYYY/MM/DD] (hereinafter, "Execution Day"), between [Company Name], a company incorporated in [jurisdiction], with [company code of XXXXX]and its principal place of business at [address] (Hereinafter "Supplier") and [Company Name], a company incorporated in [jurisdiction], with [company code of XXXXX]and its principal place of business at [address] (Hereinafter "Distributor").

本經銷協議（以下簡稱「本協議」）訂立於〔年/月/日〕（以下簡稱「簽署日」），當事人為〔公司名稱〕，係依據〔法域〕之法律設立，〔公司編號為xxxx〕，且主營業所設於〔地址〕（以下簡稱「供應商」），以及〔公司名稱〕，係依據〔法域〕之法律設立，〔公司編號為xxxx〕，且主營業所設於〔地址〕（以下簡稱「經銷商」）。

WHEREAS:

(a) Supplier designs and develops quality [products description] and related products throughout the world under the Supplier name and owns or controls the rights to use and to authorize others to use the Supplier trademark, service mark, and trade name and other intellectual property rights in connection with the design, manufacture, marketing, distribution, and/or sale of said products, together with the goodwill symbolized thereby and the business appertaining thereto; and

(b) Distributor wishes to have the right to distribute said products in the Territory of this Agreement.

NOW, THEREFORE, in consideration of the mutual promises contained herein, the parties agree as follows:

鑒於：

(a)供應商在全世界以供應商之名稱設計並開發高品質的〔產品描述〕與相關產品，並具有或控制使用權且得授權他人利用供應商之商標、服務標章與商業名稱，以及其他有關前述產品設計、製造、行銷、經銷及／或銷售之權利，以及依此象徵化的商譽，以及附屬的業務；以及

(b) 經銷商期望有權在本協議指定區域經銷前述產品。

為此，以本協議包含的相互間允諾作為約約，各方當事人合意如下：

1. Definitions - As used herein:

1.1　"Territory" shall mean [description of territory]

1.2　"Products" shall mean only [description of brand] branded footwear and related products as Supplier in consultation with Distributor may from time to time authorize Distributor in writing to sell in the Territory.

1.3　"First Cost" shall mean the total cost per unit as invoiced by the factory that manufactures [product description] or related products, including additional amounts for unamortized tooling costs and agent commissions, if any.

1.4　"Trademarks" shall mean the [description of trademark] trademark, together with all other Trademarks, service marks, trade names, style or model names, logos, copyrights, designs and other intellectual property rights owned or controlled by Supplier.

1. 定義 - 本協議使用詞語定義如下：

1.1　「指定區域」應指〔區域描述〕。

1.2　「契約產品」應專指〔品牌描述〕品牌產品，供應商得徵詢經銷商意見後隨時以書面授權經銷商在指定區域內銷售。

1.3　「最初成本」應指製造〔產品描述〕或相關產品的工廠請款的每單位總成本，如有時，亦包括未攤銷的工具成本與代理人佣金。

1.4　「產品商標」應指〔商標描述〕之商標，以及其他各項商標、服務標章、商業名稱、風格或型名、標誌、著作權、設計、與供應商所有或控制的其他智慧財產權。

2. Appointment - Subject to the terms and conditions of this Agreement

Supplier hereby appoints Distributor as its exclusive Distributor of Products in the Territory for the term of this Agreement. While Supplier cannot guarantee that Products will not enter the Territory through parallel channels, Supplier will make commercially reasonable efforts to ensure that none of their Distributors or sales outlets sell into the Territory.

2. 委任 – 在本協議的條款與條件限制下，供應商特此委任經銷商在本協議之效期內擔任指定區域內契約產品的專屬經銷商。供應商雖不能擔保契約產品不會藉由平行通路進入指定區域，供應商將盡商業上合理努力以確保其精銷商或銷售通路不會在指定區域內銷售。

3. Term of Agreement

Unless sooner terminated by either party in accordance with the provisions of this Agreement, the initial term of this Agreement shall be for a period 3 years starting on Execution Day and ending on [MM/DD/YYYY].

3. 協議效期

除非任一方當事人依據本協議之條文提早終止外，本協議之初始效期應為簽約日起並至〔年/月/日〕止。

4. Obligations and Rights of Distributor

4.1   Best Efforts - During the term of this Agreement, Distributor shall use its best efforts to:

4.1.1 Promote, develop the market for, sell and distribute Products throughout the Territory,

4.1.2 Support and cooperate in the execution of global marketing plans and strategies,

4.1.3 Select dealers and maintain facilities for the sale of

Products, and maintain a business and sales organization adequate to work and develop the Territory,

4.1.4 Confine retail and wholesale sales of Products in the Territory to (a) persons who operate suitable retail locations, as determined by Supplier in its sole discretion, and (b) wholesalers who sell to persons who operate suitable retail operations,

4.1.5 Comply with all local laws, rules, regulations and other governmental requirements.

4.2　Purchases - Except as may be agreed in writing by Supplier, Distributor shall purchase Products exclusively from Supplier.

4.3 Payments for Products - During the term of this Agreement Distributor shall pay Supplier according to price lists denominated in U.S. Dollars provided by Supplier to Distributor. Prices shall be those set forth on price lists in effect when the order is placed. Distributor shall be responsible for all freight and duties from the factory to final destination.

4.3. A Factory direct costs. The Distributor agrees to buy inventory from designated factories on a seasonal basis based on a First Cost price, plus a 20％ royalty.

4.3 B The Distributor would also purchase samples from Supplier as necessary at First Cost.

4.4 Terms and Conditions of Payment for Products

4.4.1 The supply of Products by Supplier shall be subject to Supplier's standard terms and conditions of sale as promulgated from time to time by Supplier. Each order and acceptance of such order for Products shall constitute a separate contract between Supplier and Distributor subject to the terms and conditions thereof.

4.4.2 Payment for Products shall be by wire transfer to the account designated by the Supplier, and any fees and expenses charged for the wiring shall be borne by the Distributor.

4.5 Orders and Cancellations - All orders for Products shall be submitted on forms prescribed by and in accordance with such arrangements as are advised from time to time by Supplier. Unless agreed to in writing by Supplier orders received from Distributor shall not be subject to cancellation, change or modifications by Distributor.

4.6 Records and Inspections - Complete and accurate books of account and records of Products purchased and sold by Distributor shall be maintained and retained at Distributor's offices. These records shall be available for inspection by Supplier or its authorized representative, at any reasonable time while this Agreement remains in effect and for a period of one (1) year thereafter. The right to inspect shall include the right to copy part or all of such accounts and records.

4.7 Minimum Purchases - Distributor shall use its best efforts to sell sufficient quantities of Products to maintain a level of performance to be agreed annually by Distributor and Supplier. As an absolute minimum Distributor must each year purchase from Supplier Products that are equal to or exceed the amounts indicated on Exhibit A, attached hereto.

4.8 Confidentiality Obligations-Both during and after the term of this Agreement, Distributor shall take all reasonable and practicable steps to maintain in the strictest confidence all Proprietary Information and other confidential information provided to Distributor by or through Supplier and/or its affiliated companies. The term "Proprietary Information" includes without limitation the following items: product, materials, and components research; designs; drawings; blueprints; specifications; sample requests; prototypes; models; development samples; confirmation samples; test results; inventions; discoveries; trade secrets; know-how; patent, design, copyright, and trademark applications; product briefs; patterns; molds; manufacturing methods and processes; market research reports; marketing plans and forecasts; customer list; style lists; and price schedules. This confidentiality requirement extends to all forms and materials in which Proprietary Information may be contained, including without limitation all draft and final originals, copies, memoranda; notes; reports; writings; drawings; blueprints; graphs; charts; film; fiche; photographs; tapes; discs; and other documentary electronic, or magnetic data compilations. This requirement shall not apply with respect to information which comes into the public domain other than by disclosure by Distributor.

4.9 Advertising and Promotion

4.9.1 Hereby, Supplier grants Distributor permission to reproduce the Trademarks on its advertising, promotional and marketing materials, provided that all such advertising, promotional and marketing materials used by Distributor must conform to all advertising and trademark use guidelines provided by Supplier.

4.9.2 During each calendar year of this Agreement, Distributor agrees to spend on advertising and promotions in the Territory a sum equal to not less than two percent (2%) of its total net invoiced sales of Products. As a guideline, at least two thirds of this expenditure shall be in the form of media (print, radio and/or television) advertising.

4.10 Territorial Limits - Distributor shall not, during the Term of this Agreement, advertise, promote or market Products, nor maintain agents, branch office or distribution facilities to sell Products outside the Territory. Except at the specific written request of Supplier, Distributor shall not solicit orders for Products from persons or companies outside the Territory.

4.11 Returns - Distributor agrees to accept defective product returns from its customers. Supplier shall be responsible for defective product returns deemed of substandard quality or not made to proper size and fit standards as agreed on by both parties.

4.12 Product Liability Insurance. Supplier shall obtain product liability insurance in the Territory in an amount which is commercially reasonable given the quantity of Products to be distributed under this Agreement and the potential for monetary recovery, and shall name Distributor and any successors thereof as additional insured's under such policy.

4. 經銷商的義務與權利

4.1 盡最大努力－在本協議效期內，經銷商應盡最大努力辦理以下事項：

4.1.1 推廣、開發市場以在整個指定區域內銷售並經銷契約產品，

4.1.2 支援並合作執行全球行銷計劃與策略，

4.1.3 為契約產品之銷售選擇分銷商並維持相關設施，並維持可合適運作並開發指定區域的業務與銷售組織，

4.1.4 在指定區域內分派零售與批發銷售責任給(a)經銷商全權判斷適合的零售點經營者，以及(b)銷售給經營適當零售點的批發商，以及

4.1.5 遵守本地法律、規則、法規與政府其他要求。

4.2 採購－除非供應商書面同意外，經銷商應僅向供應商採購契約產品。

4.3 為契約產品付款－在本協議效期內，經銷商應依據供應商提供給經銷商以美金計價的價目表向供應商付款。價格應為下單時有效的價目表所列之價格。經銷商應負責從工廠到最後到貨地點之間的所有運費與稅賦。

4.3.A 工廠直接成本。經銷商同意基於最初成本價格加20％權利金，按季節向指定工廠購買存貨。

4.3 B 經銷商必要時也將按最初成本向供應商採購樣品。

4.4 契約產品付款條件

　　4.4.1 供應商供應產品應受供應商隨時訂定的標準銷售條款條件之拘束。此種契約產品訂單的下單與承諾均構成供應商與經銷商之間的個別契約，而應受其相關條款與條件之拘束。

　　4.4.2 契約產品之付款應以電匯付款至供應商指定帳戶方式為之，且匯款時收取的各種費用與款項均應由經銷商負擔。

4.5 訂單與解除 - 所有契約產品之訂單應按照供應商隨時指定之表單，並依據供應商隨時指示之安排發送。除非供應商另有書面同意，經銷商對於其發送的訂單不得解除、變更或修改。

4.6 記錄與檢查 – 經銷商之辦公室應維持並保留完整且準確的會計帳冊與記錄以記載經銷商購買並銷售的契約產品。在本協議效期內與失效後一年內供應商或其授權的代表得在任何合理的時間檢查前述記錄。檢查權應包括複製部分或全部的帳冊與記錄。

4.7 最小購買量 – 經銷商應盡最大努力銷售數量足夠的契約產品以維持經銷商與供應商每年合意決定的履約水準。經銷商每年應向供應商購買的契約產品最小數量應大於本協議附件A所列之數額。

4.8 保密義務 - 無論本協議效期內或效期後，經銷商均應採行各種合理且實際的措施以最嚴格的程度維護供應商及／或其關係企業提供給經銷商的各種私有資訊或其他機密資訊。所稱「私有資訊」包括但不限於

以下各項：產品、材料與元件的研究、設計、圖樣、藍圖、規格、樣品需求、原型、模型、開發樣本、確認樣本、測試結果、發明、發現、營業秘密、技術訣竅、專利、設計、著作權與產品商標申請、產品簡報、模型、模具、製造方法與流程、市場研究報告、行銷計畫與預測、客戶清單、程式表、風格清單與價目表。本項保密要求擴及可能含有私有資訊的各種表單與材料，包括但不限於各種草稿與最終稿本、副本、備忘錄、筆記、報告、寫作、繪圖、藍圖、圖示、圖表、電影、微縮膠片、照片、錄音帶、光碟與其他電子記錄媒體，或磁性資料檔案集合。本要求不適用於非因經銷商之揭露而已進入公共領域的資訊。

4.9 廣告與促銷

4.9.1 供應商茲許可經銷商得在廣告、促銷與行銷材料中重製產品商標，但經銷商採用的所有此等廣告、促銷與行銷材料必須符合供應商提供的各項廣告與產品商標使用準則。

4.9.2 本協議效期內每一日曆年，經銷商同意在指定區域內投入不少於相關契約產品銷售淨額百分之二的廣告與促銷活動費用。原則上，此項支出至少三分之二應投入媒體形式的廣告（出版品、廣播及／或電視）。

4.10 區域限制 – 在本協議效期內，經銷商不得在指定區域外廣告、促銷或行銷契約產品，亦不得委任代理人、分公司經銷設施在指定區域外銷售契約產品。除供應商以書面具體要求外，經銷商不得向指定區域外的個人或公司招徠訂單。

4.11 退貨 – 經銷商同意接受客戶退回瑕疵產品。供應商同意負責退回品質標準低劣、不符適當尺寸，不符雙方合意合身標準的產品。

4.12 產品責任保險。供應商應在指定區域內取得產品責任保險，且

應依據本協議將經銷的契約產品數量與潛在的金錢賠償金額決定商業上合理的保險金額，且應將經銷商與其任何繼受人也列為此等保單的被保險人。

5. Supplier's Retail Operations - This Agreement does not authorize Distributor to operate itself or grant to others the right to operate Supplier retail stores (i.e., those bearing the name of Supplier as part of the store name or in which the store appears to be owned or controlled by Supplier). The operation of any such Supplier retail sales operations shall require advance written authorization by Supplier. The operation of any Supplier retail store by any third party must be subject to separate agreement between Supplier and Distributor.

5. 供應商之零售營運-本協議並未授權經銷商自行或授權其他人經營供應商零售店之權利（亦即以供應商之名稱作為商店名稱之一部分，或者商店顯示供應商所有或控制之外觀）。此等供應商零售經營需要取得供應商事先書面同意。任何第三人如擬經營任何供應商零售店必須與供應商及經銷商另行簽訂契約。

6. Use Supplier's Name as Company or Trading Name - The name of Supplier shall not be used as the trading or company name of Distributor without the written consent of Supplier. Nothing contained herein, however, shall be construed to preclude Distributor from representing itself on its letterhead and other stationery, etc as Supplier's exclusive Distributor in the Territory, or from using the Supplier trademark and logo on its stationery and advertising, provided that the trading or company name of the Distributor also appears on such stationery and advertising and provided further that

such is done in a manner approved by Supplier.

6. 使用供應商之名稱作為公司名稱或商業名稱－經銷商未經供應商書面同意不得使用供應商之名稱作為經銷商之商業名稱或公司名稱。然而，本協議不得解釋為禁止經銷商在任何信函表頭與其他文宣物品上標明為經銷商在指定區域內的專屬經銷商，或禁止在其文宣物品與廣告物品上使用供應商之商標與標誌，但經銷商之商業名稱或公司名稱亦應同時出現在相同的文宣物品與廣告物品上，而且應按照供應商許可之方式為之。

7. Trademarks

7.1 Distributor acknowledges that the Trademarks and other intellectual property have substantial goodwill.

7.2 This Agreement does not constitute and shall not be construed as a license of the Trademarks or other intellectual property. Distributor acknowledges that it does not claim any ownership rights in the Trademarks and other intellectual property and that it shall not acquire or claim any ownership rights therein by reason of this Agreement, or as a result of use, or for any other reason. Distributor will not at any time do, or knowingly permit others to do, any act or thing which would in any way impair the rights of Supplier in and to the Trademarks or which may affect the validity of the Trademarks or which may dilute or depreciate the value of the Trademarks or their reputation. Distributor agrees that any rights in the Trademarks which may arise by virtue of Distributor activities pursuant to this Agreement or by operation of law shall vest in and, at the written request of Supplier, shall be assigned to Supplier and without charge.

7.3 Distributor shall not affix any other trademark or logo onto the Products or any packaging for the Products.

7.4 Distributor undertakes not to copy, produce, make, modify or manufacture or assist any person to copy, produce, make or manufacture goods similar to Products or any part thereof for use, sale or any other purpose.

7.5 Distributor acknowledges the sole right of Supplier to file and prosecute any trademark or other application relating to the Trademarks as Supplier may deem advisable including any such applications arising from or made necessary by the activities of the Distributor under this Agreement. Distributor will, when requested by Supplier, cooperate with Supplier in connection with any such applications, including applications to record the Distributor as a registered or permitted user under new or existing trademark registrations or applications. The expenses of preparing and conducting any applications or registrations and the filing of any registered user agreements shall be borne by Supplier.

7.6 Distributor shall, upon the expiration of ninety (90) days from the termination of this Agreement, cease and desist from all use of the Trademarks and other intellectual property in any way and shall deliver up to Supplier or destroy in a manner approved by Supplier, all material, signage, documents and papers upon which the Trademarks appear.

7.7 Distributor shall comply with Supplier's guidelines on proper usage of the Trademarks and with all applicable laws with respect to printing and placement of proper notice of the Trademarks on Products, and on all packaging and advertising, marketing or promotional material. All such materials shall indicate that the Trademarks are owned or controlled by Supplier, and have been applied by or under license from Supplier as is appropriate in each case.

7.8 Distributor shall:

7.8.1 make regular checks within the Territory to see if there is any infringement or other violation of the Trademarks or any other rights of Supplier, including any unauthorized use of the Trademarks within the Territory (hereinafter referred to as "Illegal Acts"),

7.8.2 Promptly inform Supplier in writing of all Illegal Acts of which it becomes aware.

7.8.3 On request, provide Supplier with as much information as Distributor can reasonably obtain about all such Illegal Acts.

7.8.4 Upon learning of any Illegal Acts, Supplier shall be entitled, at its discretion, to take such action as it considers necessary or appropriate to enforce Supplier's rights, including without limitation action to suppress or eliminate the Illegal Acts. Supplier shall also be entitled to seek recovery for all damages resulting there from. Distributor shall have no authority to enforce the rights of Supplier in and to the Trademarks and other intellectual property, nor shall Distributor have any control over action taken by Supplier to enforce such rights.

7.8.5 Distributor shall, at the request of Supplier, make available to Supplier free of charge all information and particulars in its possession which will assist Supplier to deal with Illegal Acts and will, at Supplier's request and expense, join in any action necessary to suppress and prevent any Illegal Acts or the continuation thereof.

7.8.6 Supplier shall be entitled to all damages, costs and other sums which may be due or recoverable as a result of any Illegal Acts.

7. 產品商標

　　7.1 經銷商確認產品商標及其他智慧財產權具有重大的商譽價值。

　　7.2 本協議並無產品商標授權或其他智慧財產權之授權，並不得解釋為此等授權。經銷商確認其對商標與其他智慧財產權不得主張任何所有權，而且亦未因為本協議或因為使用或其他理由而獲得或主張其任何所有權。經銷商任何時候均不得，亦不得在知情之情況下容許其他人為任何行為或作任何事而可能損害供應商對產品商標之權利或可能影響商標之有效性或可能淡化或減損商標或其聲譽之價值。經銷商同意因為經銷商依據協議進行之活動或因為法律的適用而可能產生對產品商標之各項權利，在供應商書面請求時均應無償移轉於供應商。

7.3 經銷商不得在契約產品或契約產品的任何包裝上附加任何其他商標或標誌。

7.4 經銷商承諾不得為了使用銷售或其他任何目的複製、重製、製作、修改或製造或協助任何人複製、重製、製作、修改類似於契約產品之商品，或其任何部分。

7.5 經銷商確認僅供應商有權進行供應商認為適當的產品商標相關商標申請，或應用申請，或追究商標侵權責任，包括因為經銷商依據本協議進行之活動而可能產生或屬必要的任何應用。在供應商要求時，經銷商將與供應商就有關此等申請之事項進行合作，包括在新的或既有的商標登記或申請案中申請將經銷商記錄為登記使用者或許可之使用者。準備或辦理任何申請或登記以及申請任何登記使用者契約之費用均應由供應商負擔。

7.6 經銷商在本協議終止後九十（90）日內應停止並結束任何產品商標與其他智慧財產權任何方式之使用，並應將顯示產品商標的所有材料、標誌、文書與紙製品交付供應商或以供應商許可之方式銷毀。

7.7 經銷商應遵守供應商制訂的產品商標適當使用之準則，並遵守有關契約產品上適當的產品商標告知之印製與顯示，以及有關各種標準與廣告、行銷或促銷材料之應適用法律。所有此等材料均應標明產品商標由供應商所有或控管，也已由供應商自行申請或依供應商之授權而申請。

7.8 經銷商應：

7.8.1 在指定區域內定期檢查有無侵害或其他違反產品商標或供應商任何其他權利之情事，包括未經許可在指定區域內使用產品商標（以下簡稱「非法行為」）。

7.8.2 如果知悉有任何非法行為，應立即以書面告知供應商。

7.8.3 依供應商之要求，提供經銷商可取得而有關此等非法行為之各種資訊。

7.8.4 知悉各種非法行為後，供應商應有權依其裁量採取其所認為必要或適當的各種行動以施行供應商之權利，包括但不限於遏止或消除非法行為。供應商也應有權請求賠償因此所生之任何損害。經銷商無權強制實施供對產品商標與其他智慧財產之權利，經銷商對於供應商採行實施此等權利之行為亦無任何控制權。

7.8.5 在供應商提出請求時，經銷商應無償向供應商提供經銷商所持有可輔助供應商處理非法行為的各種資訊，且在供應商請求並負擔費用時，應加入為遏止及預防任何非法行為，或予以終止而屬必要的任何行動。

7.8.6 供應商應有權請求因為任何非法行為而發生或可求償的任何損害、成本與其他金額。

8. Special Rights of Termination

8.1 If either party shall actually or effectively cease to conduct its business; or shall make any involuntary assignment of either its assets or its business for the benefit of creditors; or if a trustee or receiver or administrator is appointed to administer or conduct its business affairs; or if any insolvency, bankruptcy or similar proceedings are commenced by itself or against it, then the other party may terminate this Agreement with immediate effect upon written notice to such party.

8.2 If Distributor fails to remit to Supplier within sixty (60) days from the time specified as the due date any sum payable to Supplier under the terms of this Agreement, including payment for Products, Supplier may terminate this Agreement by providing thirty (30) days written notice to Distributor.

8.3 Supplier may terminate this Agreement with immediate effect upon written notice to Distributor in the event that:

8.3.1 Distributor violates the territorial limits prescribed herein; or

8.3.2 Distributor fails to meet the purchase minimums set out in Paragraph 4.8;

8.3.3 The control of Distributor shall pass from the present shareholders to other persons whom Supplier shall in its absolute discretion regard as unsuitable; or

8.3.4 Distributor or any of its directors or officers acts in manner which in the opinion of Supplier is likely to bring Supplier into disrepute.

8. 特別終止權

8.1 如果任一方當事人事實上或法律上停止營業，或自願地為債權人之利益移轉其資產或事業；或委任受託人、管理人、檢查人已管理或經營其事業，或如果自行聲請或被聲請任何無力償債、破產或類似程序，則他方當事人得以書面通知該當事人立即終止本協議。

8.2 如果經銷商未在本協議條款規定應付供應商款項之日六十（60）日內匯付供應商，包括契約產品之付款，供應商得於三十（30）

日前以書面通知經銷商終止本協議。

8.3 如有下列情事，供應商得立即以書面通知經銷商終止本協議：

8.3.1 經銷商違反本協議規定的區域限制；或

8.3.2 經銷商未符合第4.8條規定的最低採購數量要求；

8.3.3 經銷商之控制權由目前的股東移轉於其他人，而依供應商之全權裁量認為前述受讓人不適當者；或

8.3.4 經銷商或其任何董事或主管以供應商認為可能損害供應商名譽之方式行事。

9. Termination on Default - In addition to the special rights of termination provided in the preceding paragraph, in the event that Supplier or Distributor shall fail to perform any of their obligations to the other party under this Agreement, and such non-performance is not cured within thirty (30) days after written notice of said non-performance being given by the other party, then the other party may give written notice to the defaulting party to terminate this Agreement with immediate effect.

9. 因違約而終止-除前條所定特別終止權之外，如果供應商或經銷商未依據本協議履行對他方當事人之任何義務，且此等不履行在他方當事人書面通知不履行之事後三十（30）日內未改善者，則他方當事人得以書面通知違約方本協議並立即生效。

10. Rights and Duties on Expiration or Termination of Agreement

10.1 In the event of the expiration or termination of this Agreement for any reason:

10.1.1 Distributor shall immediately cease holding itself out to third parties as being associated with Supplier.

10.1.2 Distributor shall return all Proprietary Information and other confidential information previously received from Supplier, in whatsoever form contained, which is in Distributor's possession or control, including all originals, copies, reprints, translations and samples thereof, within twenty (20) days after receipt of a written request from Supplier. In addition, Distributor shall provide Supplier within twenty (20) days with a list of all trade customers who have purchased Products from Distributor within the previous 12 months, including the addresses of such customers.

10.1.3 Outstanding unpaid invoices between Supplier and Distributor shall become immediately payable.

10.1.4 Distributor shall have no further rights to use the Trademarks or any of Supplier's intellectual property rights and in particular but without prejudice to the generality of the foregoing shall cease to use the Trademark on its letterheads, packaging or elsewhere.

10.1.5 Distributor shall immediately cease purchasing, selling, advertising, and/or distributing any Products, except as provided for herein. Distributor shall not be entitled to delivery of any then outstanding Supplier orders, whether or not such orders have been accepted by Supplier.

10.1.6 Distributor shall for 120 (one hundred and twenty) days following the termination date have the right to fulfill from its inventory any orders outstanding for Products at the date of termination. Distributor shall furnish Supplier with a list of all outstanding orders within twenty days of the termination date.

10.1.7 As to any remaining inventory, Supplier shall have the option to buy the whole or any part of said inventory from Distributor at Distributor purchase cost (including duties, taxes and delivery charges) or wholesale market value, whichever is lower.

10.1.8 If Supplier fails to exercise the option set forth in subparagraph 10.1.7 above by the end of the one hundred and twenty (120) day sell-off period provided in subparagraph 10.1.6 above, Distributor shall be free to sell any remaining inventory.

10.2 Supplier or its designated representatives shall have the right during the one hundred and twenty (120) day period prior to the expiration of the term of this Agreement or its termination to advertise Products, approach customers and to take orders for the delivery of Products in the Territory after the expiration date.

10.3 Termination of this Agreement shall not prejudice any rights of cither party which have arisen on or before the date of termination, and which are intended to have continuing effect.

10. 本協議屆期或終止後的權利與義務

　10.1 本協議因任何理由而屆期或終止時：

　　10.1.1 經銷商應立即停止對外表現或顯示與供應商有任何關係。

　　10.1.2 經銷商應在收到供應商書面請求後二十（20）日內返還之前由供應商收到而由經銷商持有或控管的各種私有資訊與其他資訊，無論提供資訊的形式為何，包括各種原件、影本、複印本、翻譯與樣本。此外，經銷商應在二十（20）日內提供前十二個月內曾向經銷商購買契約產品的各個交易顧客之名單，包括此等顧客的地址。

　　10.1.3 供應商與經銷商之間未付的款項應立即支付。

　　10.1.4 經銷商無權再使用產品商標或供應商的智慧財產權，而且在不妨礙前述約定之普遍效力下，經銷商應立即停止在信函表頭、包裝或其他地方使用產品商標。

　　10.1.5 除本協議另有約定外。經銷商應立即停止採購、銷售、廣告及/或經銷任何契約產品。

　　10.1.6 經銷商在終止日後120（一百二十）日內有權以其庫存履行終止時未完成之契約產品訂單。經銷商應在終止日起二十日內向供應商提出所有未完成訂單的清單。

　　10.1.7 關於剩下的庫存，供應商有權按照採購成本（包括稅賦與運費）或市場批發價（以較低者為準）向經銷商購回前述庫存之全部或任何部分。

10.1.8 如果供應商未在前述10.1.6規定的一百二十（120）日清倉期結束時行使其權利，經銷商得自主出售任何剩下的庫存。

10.2 供應商或其指定代表人在本協議屆期或終止的一百二十（120）日前得在指定區域，契約對契約產品進行廣告、接觸客戶、接受訂單，並在屆期後才交付契約產品。

10.3 本協議之終止不妨礙在終止日當日或之前已發生的任何權利，且該權利應繼續有效。

11. Indemnity

11.1 Supplier shall indemnify and hold Distributor harmless from any and all claims, actions and demands made or brought by any third party claiming that the Distributor use of the Trademarks infringes or otherwise violates any rights of such third party, provided that such claim or demand does not arise as a result of any action by Distributor which is not expressly or implicitly permitted by this Agreement. Distributor shall inform Supplier immediately of any such claim and make no admission in relation thereto. Supplier may, at its own expense, be entitled to conduct and or settle all negotiations and litigation so arising and Distributor agrees to be bound by any settlement or agreement reached by Supplier in such matter. Distributor shall assist Supplier in preparation of the defense of any such claim, action or demand by providing whatever information or assistance from its staff as may reasonably be required.

11.2 Distributor shall indemnify and hold Supplier harmless from claims, actions and demands made or brought by any third party in connection with and arising out of Distributor activities under this Agreement. Distributor shall vigorously defend any legal action brought against it in connection with its activities under this Agreement and engage counsel approved by Supplier.

11. 補償

11.1 供應商應補償經銷商並使其不因第三人主張經銷商使用產品商標侵害或以其他方式違反該第三人之任何權利而提出之任何一切求償、行動與要求而受到任何損害，但前述求償或要求不得起因於非屬本協議明示或默示許可經銷商所為之行為。經銷商應立即告知供應商前述求償且不得對之為任何認諾。供應商得自行負擔費用進行談判與訴訟和解，且經銷商同意受供應商就此事項達成之和解或安排之拘束。經銷商應協助供應商藉由依據供應商之員工可能合理要求的提出任何資訊或協助而準備任何此等訴訟、行動或要求之答辯。

11.2 經銷商應補償並使供應商不因為有關於或起因於經銷商依據本協議之活動第三人提出任何求償、行動與要求而受到任何損害。經銷商應積極地辯護與本協議有關而針對供應商提出的任何法律行動，並委託供應商許可的律師。

12. Miscellaneous.

12.1 This Agreement is non-assignable in whole or in part by Distributor without the written consent of Supplier. Transfer of the ownership of the assets or the shares of Distributor from the shareholders at the date of execution of this Agreement shall, for the purposes of this paragraph, be considered an assignment by

Distributor requiring Supplier's written consent. Distributor shall at Supplier's request provide Supplier such information as it shall reasonably require to verify the ownership and control of Distributor.

12.2 Supplier and Distributor shall each execute and deliver all such instruments and do such acts as may be necessary or reasonably required by the other party to evidence or give effect to this Agreement or its terms.

12.3 Supplier and Distributor have entered into this Agreement as independent contractors only. This Agreement does not constitute Distributor as the agent or legal representative of Supplier nor does it constitute Supplier as the legal representative of Distributor. Neither party shall have any right or authority to assume or create any obligation or responsibility, express or implied, on behalf of or in the name of the other, or to bind the other in any manner. Furthermore, whenever Distributor describes its relationship to Supplier it shall make it clear that it is a Distributor and not an agent or representative and that it has no authority to make contracts or incur obligations binding on Supplier.

12.4 Failure of a party to enforce one or more of the provisions of this Agreement or to exercise any option or other rights hereunder or to require at any time performance of any of the obligations hereof shall not be construed to be a waiver of such provisions by such party or in any way to affect the validity of this Agreement or such party's right thereafter to enforce each and every provision of this Agreement, nor to preclude such party from taking any other action at any time which it would legally be entitled to take.

12. 雜項條款

　　12.1 未經供應商之同意，經銷商不得將本協議之全部或一部轉讓他人。本協議簽署時之經銷商股東嗣後將經銷商之資產或股份轉讓他人時，應視為本項所稱之轉讓而須徵得供應商之書面同意。在供應商要求時，經銷商應提供供應商認為合理必要的資訊以確認經銷商之所有權與控制關係。

　　12.2 供應商與經銷商應簽署並交付他方為使本協議或其條款發生效力而需要或合理要求的文書。

　　12.3 供應商與經銷商僅以獨立包商之身份締結本協議。本協議不使經銷商成為供應商之代理人或法定代表，亦不使供應商成為經銷商之法定代表。任一方當事人均無任何權利或權限明示或默示代表他方或以他方名義承擔或創設任何義務或責任，或使他方以任何方式受拘束。而且，任何時候經銷商說明其與供應商之關係時，經銷商應表明其為經銷商，且非代理人或代表人，且亦無權簽署契約或承擔義務而拘束供應商。

　　12.4 一方當事人未強制履行本協議之一項或多項條款，或未行使任何本協議之任何選擇或其他權利，或在任何時候未要求他方當事人履行本協議之任何義務，不得以此解釋為該方當事人拋棄依該等條款享有之權利，或以任何方式影響本協議之效力，亦不排除該方當事人在任何時候依其享有之權利採取其他行動。

13. Notices - All notices and requests for written approval provided for herein, shall be submitted by facsimile transmission, courier or airmail as appropriate. Additionally, all formal notices to Supplier,

annual accounts, audit certificates etc. shall be sent by certified or registered mail, return receipt requested, or by hand delivery to the Supplier at the address noted on the first page of this Agreement. All notices to Distributor shall be sent to the address noted on the first page of this Agreement. Either party to this Agreement may provide the other party with written notices of changes of address to be thereafter used for the purpose of this Agreement.

13. 通知－所有依據本協議所為之通知或請求書面許可，均應以傳真、快遞或航空郵件擇其適當者為之。此外，對供應商寄發的各項正視通知、年度會計帳冊、稽核憑證等，均應以認證或掛號郵件、附回郵，或專人親自送達在本協議首頁記載的供應商地址。所有發給經銷商的通知應寄到本協議首頁記載的地址。本協議任一方當事人得以書面通知他方變更地址，並於通知後適用於本協議。

14. Separability - If any provision of this Agreement is held void by a final judgment or decree of any court, commission or other judicial or quasi-judicial body of competent jurisdiction, this Agreement as a whole shall remain in force and effect in all other respects as if said provisions had not been included in this Agreement, unless said judgment of invalidity affects the contract as a whole.

14. 效力可分－如果本協議任何條款在有管轄權的任何法院、委員會、司法機關或準司法機關之終局判決或裁定中被認定無效，本協議其他部分仍應有效，視同本協議並不包括前述條款，除非前述條款無效之判決影響整個契約之效力。

15. Force Majeure - No party shall be liable for the failure to carry out its obligations hereunder in the event that it is prevented from doing

so by war, unavailability of shipping vessels, insurrection, governmental action prohibiting importation of goods or any other similar causes beyond the control of the party.

15. 不可抗力 - 如果因為戰爭、無船舶可用、暴亂、禁止商品進口之政府行動，或其他超出當事人控制範圍之類似原因，致無法履行義務，當事人為此不承擔責任。

16. Governing Law/Jurisdiction

Any dispute arising out of or in connection with this Agreement, including any question regarding its existence, validity or termination, shall be referred to and finally resolved by arbitration administered by the Singapore International Arbitration Centre ("SIAC") in accordance with the Arbitration Rules of the Singapore International Arbitration Centre ("SIAC Rules") for the time being in force, which rules are deemed to be incorporated by reference in this clause. The seat of the arbitration shall be Singapore. The Tribunal shall consist of one (1) arbitrator. The language of the arbitration shall be English. This Agreement shall be governed by laws of Singapore and construed accordingly.

16. 準據法 / 管轄。

任何起因於或有關於本協議之糾紛，包括任何有關其存在、效力、或終止者，均應提交新加坡國際仲裁中心（「SIAC」）依據當時有效的新加坡國際仲裁中心仲裁規則（「SIAC 規則」）進行仲裁並最終解決，該規則應視同已納入本條。仲裁地應為新加坡。仲裁庭應由一 (1) 位仲裁人組成。仲裁語言應為英文。本協議應以新加坡法為準據法，並依此

解釋。

17. Entire Agreement - This Agreement sets forth entire understanding and agreement and supersedes with effect from the commencement of this Agreement any and all prior understandings, contracts or agreements between Distributor and Supplier with respect to the subject matter of this Agreement. There are no representations or promises between the parties hereto except as set forth herein.

17. 完整合意 – 本協議列出經銷商與供應商之間有關本協議主旨事項之完整瞭解與合意,並自本協議生效時取代渠等之間其他任何先前之瞭解、契約或合意。除本協議明訂之聲明或允諾,當事人間並無其他任何聲明或允諾。非於嗣後由雙本協議任何條款

[SIGNATURES FOLLOW ON THE NEXT PAGE]

[簽名頁在次頁]

IN WITNESS WHEREOF, the Parties have duly authorized and caused this Distribution Agreement to be executed as follows:

鑒於上述事項，各方當事人茲授權並促使以下人適簽署本經銷協議：

The Supplier

供應商

[Company Name]

[公司名稱]

By_____

簽約人

　　[Signature]

　　〔簽名〕

Name_____

姓名

　　[Print]

　　[正楷]

Title

職銜

The Distributor

經銷商

[Company Name]

[公司名稱]

By_____

簽約人

　　[Signature]

　　〔簽名〕

Name_____

姓名

　　[Print]

　　[正楷]

Title

職銜

Exhibit A: Minimum Purchasement

附件A:**最低購買量**

20xx:

20xy:

20xz:

國家圖書館出版品預行編目資料

律師教你寫英文契約：專業人士必備的實務指南 / 高忠義 著. --
初版. -- 臺北市：商周出版，城邦文化事業股份有限公司出版：
英屬蓋曼群島商家庭傳媒股份有限公司城邦分公司發行，
2024.03
面； 公分
ISBN 978-626-390-051-6（平裝）
1. CST: 商業英文　2. CST: 商業應用文　3. CST: 契約
493.6　　　　　　　　　　　　　　　　　　113001978

律師教你寫英文契約：
專業人士必備的實務指南

作　　　　者／高忠義
責 任 編 輯／林瑾俐

版　　　　權／吳亭儀
行 銷 業 務／周丹蘋、賴正祐
總　 編　 輯／楊如玉
總　 經　 理／彭之琬
事業群總經理／黃淑貞
發　 行　 人／何飛鵬
法 律 顧 問／元禾法律事務所　王子文律師
出　　　　版／商周出版
　　　　　　　城邦文化事業股份有限公司
　　　　　　　台北市南港區昆陽街16號4樓
　　　　　　　電話：(02) 2500-7008　傳眞：(02) 2500-7579
　　　　　　　E-mail：bwp.service@cite.com.tw
發　　　　行／英屬蓋曼群島商家庭傳媒股份有限公司城邦分公司
　　　　　　　台北市南港區昆陽街16號5樓
　　　　　　　書虫客服服務專線：(02) 2500-7718 · (02) 2500-7719
　　　　　　　服務時間：週一至週五09:30-12:00 · 13:30-17:00
　　　　　　　24小時傳眞服務：(02) 2500-1990 · (02) 2500-1991
　　　　　　　郵撥帳號：19863813　戶名：書虫股份有限公司
　　　　　　　E-mail：service@readingclub.com.tw
　　　　　　　歡迎光臨城邦讀書花園 網址：www.cite.com.tw
香 港 發 行 所／城邦（香港）出版集團有限公司
　　　　　　　香港九龍土瓜灣土瓜灣道86號順聯工業大廈6樓A室
　　　　　　　電話：(852) 2508-6231　　傳眞：(852) 2578-9337
　　　　　　　E-mail：hkcite@biznetvigator.com
馬 新 發 行 所／城邦（馬新）出版集團 Cité (M) Sdn. Bhd.
　　　　　　　41, Jalan Radin Anum, Bandar Baru Sri Petaling,
　　　　　　　57000 Kuala Lumpur, Malaysia
　　　　　　　電話：(603) 9057-8822　　傳眞：(603) 9057-6622
　　　　　　　E-mail：services@cite.my

封 面 設 計／李東記
內 文 排 版／新鑫電腦排版工作室
印　　　　刷／韋懋實業有限公司
經　 銷　 商／聯合發行股份有限公司
　　　　　　　電話：(02) 2917-8022　傳眞：(02) 2911-0053
　　　　　　　地址：新北市231新店區寶橋路235巷6弄6號2樓

■ 2024年3月初版
定價 480 元

Printed in Taiwan
城邦讀書花園
www.cite.com.tw

| 廣　告　回　函 |
| 北區郵政管理登記證 |
| 台北廣字第000791號 |
| 郵資已付，免貼郵票 |

115台北市南港區昆陽街16號5樓

英屬蓋曼群島商家庭傳媒股份有限公司　城邦分公司

- -

請沿虛線對摺，謝謝！

| 書號：BJ0093 | 書名：律師教你寫英文契約：專業人士必備的實務指南 | 編碼： |

讀者回函卡

線上版讀者回函卡

感謝您購買我們出版的書籍！請費心填寫此回函卡，我們將不定期寄上城邦集團最新的出版訊息。

姓名：＿＿＿＿＿＿＿＿＿＿＿＿＿＿＿＿＿ 性別：□男 □女

生日：西元＿＿＿＿＿＿年＿＿＿＿＿月＿＿＿＿＿日

地址：＿＿＿＿＿＿＿＿＿＿＿＿＿＿＿＿＿＿＿＿＿＿

聯絡電話：＿＿＿＿＿＿＿＿＿ 傳真：＿＿＿＿＿＿＿＿＿

E-mail：

學歷：□ 1. 小學 □ 2. 國中 □ 3. 高中 □ 4. 大學 □ 5. 研究所以上

職業：□ 1. 學生 □ 2. 軍公教 □ 3. 服務 □ 4. 金融 □ 5. 製造 □ 6. 資訊

　　　□ 7. 傳播 □ 8. 自由業 □ 9. 農漁牧 □ 10. 家管 □ 11. 退休

　　　□ 12. 其他＿＿＿＿＿＿＿＿＿＿＿＿＿＿＿＿＿

您從何種方式得知本書消息？

　　　□ 1. 書店 □ 2. 網路 □ 3. 報紙 □ 4. 雜誌 □ 5. 廣播 □ 6. 電視

　　　□ 7. 親友推薦 □ 8. 其他＿＿＿＿＿＿＿＿＿＿＿＿＿

您通常以何種方式購書？

　　　□ 1. 書店 □ 2. 網路 □ 3. 傳真訂購 □ 4. 郵局劃撥 □ 5. 其他＿＿＿＿

您喜歡閱讀那些類別的書籍？

　　　□ 1. 財經商業 □ 2. 自然科學 □ 3. 歷史 □ 4. 法律 □ 5. 文學

　　　□ 6. 休閒旅遊 □ 7. 小說 □ 8. 人物傳記 □ 9. 生活、勵志 □ 10. 其他

對我們的建議：＿＿＿＿＿＿＿＿＿＿＿＿＿＿＿＿＿＿＿

＿＿＿＿＿＿＿＿＿＿＿＿＿＿＿＿＿＿＿＿＿＿＿＿＿＿

＿＿＿＿＿＿＿＿＿＿＿＿＿＿＿＿＿＿＿＿＿＿＿＿＿＿